CIRIA Report 157

1996

Guidance on the disposal of dredged material to land

J Brooke
H Paipai
S E Magenis
S L Challinor
M van Zijderveld
J Gibson
L Moore

CONSTRUCTION INDUSTRY RESEARCH AND INFORMATION ASSOCIATION
6 Storey's Gate, Westminster, London SW1P 3AU
E-mail switchboard @ ciria.org.uk
Tel 0171-222 8891   Fax 0171-222 1708

Preface

On 1 May 1994 the Waste Management Licensing Regulations came into force. The Regulations are detailed and complex, and will significantly influence the disposal of dredged material to land. Under the Regulations, unless an operator can demonstrate that the material is not controlled "waste" (i.e. it has not been discarded) or that the disposal option meets one of the specified exemptions, a Waste Management Licence will be required. This will involve operators in licence application procedures and, if appropriate, exemption registration, with associated information requirements, time commitment and cost.

CIRIA Research Project 478 Stage 2 has prepared guidance on the safe, economical and effective disposal and management of dredgings. This guidance is consistent with the aims of the Environmental Protection Act 1990 and associated Waste Management Regulations. It has identified how Government advice applies to dredgings and describes how control is to be exercised over disposal. This report will assist operators in interpreting the Regulations and applying them to their own activities in a legally correct and cost-effective manner.

The report is presented in two parts: Part A provides guidance on the selection and planning of a disposal method, including the legal position, information requirements and application procedures; Part B provides supporting information to enable the user to apply for a licence or register an exemption.

Guidance on the disposal of dredged material to land
Construction Industry Research and Information Association
Report 157, 1996

Keywords
Waste, dredgings, disposal, Waste Management Licensing Regulations, exemption.

Reader Interest
Operators and regulators concerned with dredgings disposal. Consulting engineers: waste management and disposal.

ISSN: 0305 408X
ISBN: 0 86017 450 6

© CIRIA 1996

CLASSIFICATION	
AVAILABILITY	Unrestricted
CONTENT	Guidance on the disposal of dredgings to land
STATUS	Committee guided
USER	Operators, regulators and consulting engineers concerned with dredgings disposal

Published by CIRIA, 6 Storey's Gate, Westminster, London SW1P 3AU

Foreword

This report presents the results of Research Project 478 Stage 2, *Guidance on the disposal of dredgings to land*, produced as part of CIRIA's water engineering programme. The document also constitutes NRA R&D Report 31. The object was to prepare guidance on the safe, economical and effective disposal and management of dredgings, to help operators understand the implications of the Waste Management Licensing Regulations 1994.

The report was written under contract to CIRIA by Jan Brooke, Helen Paipai and Steve Challinor of Posford Duvivier Environment, and Steve Magenis and Marcel van Zijderveld of Posford Duvivier. Chapters 3 and 8 were written by John Gibson of the Cardiff Law School and Lesley Moore of Risk Policy Analysts respectively. Contributions to the reports are gratefully acknowledged.

Steering Group

Following established CIRIA practice, the research was guided by a Steering Group which comprised:

Chairman

Mr Colin Holmes	Scott Wilson Kirkpatrick & Partners

Members

Mr Nigel Bray	National Rivers Authority - Flood Defence
Mr Neville Burt	HR Wallingford
Mr George Clapton	National Association of Waste Regulation Officers (Greater Manchester Waste Regulation Authority)
Mr Fraser Clift	Port of London Authority
Mr David Griffiths	National Rivers Authority - Pollution Control
Mr Roger Hanbury	British Waterways
Mr Ray Howells	Manchester Ship Canal Company
Ms Siân John	CIRIA
Mr Simon Mowat	Land and Water Services
Mr Nick Smith/Dr Paul Beckwith	British Waterways
Mr David Whitehead	British Ports Association

Corresponding members

Mr John Archer	ADAS
Mr Charles Ford	Charles R Ford & Associates
Mr David Gerry (until summer 1994)	Basingstoke Canal Authority
Dr Janet Gronow	Department of the Environment - Waste Technical
Mr David Noble	Association of Drainage Authorities
Ms Sue Taylor	National Association of Waste Disposal Contractors
Mr Ian Walker	Gloucester Harbour Trustees
Mr Vaughan Wilkinson	West Midlands Hazardous Waste Unit
Mr Paul Wright	Department of the Environment - Waste Management

CIRIA's Research Manager for the project was Ms Siân John. The project was initiated by Dr Judy Payne.

Project funders

The project was funded by:

British Waterways
National Rivers Authority
Manchester Ship Canal Company
Port of London Authority.

CIRIA is grateful for the support given to the project by the funders, the members of the Steering Group and all of those who were involved in the consultation undertaken as part of the project, in particular:

Central Dredging Association (CEDA)
Cheshire Waste Regulation Authority
Cornwall County Council
English Nature
Harwich Haven Authority
Health and Safety Executive
ICI Chemicals & Polymers Ltd
MAFF - Fisheries & Environmental Protection Division
South Yorkshire Waste Regulation Unit
Taylor Dredging
The Broads Authority
Tyne & Wear Waste Regulation Authority
Water Group
West Midlands Hazardous Waste Unit
Westminster Dredging Company.

Summary contents

List of figures

List of tables

List of boxes

List of abbreviations

ADAS	-	formerly known as the Agricultural Development and Advisory Service
BSI	-	British Standards Institution
BW	-	British Waterways
CADW	-	Welsh Historic Monuments
CBA	-	Cost-benefit analysis
CCW	-	Countryside Council for Wales
CDM	-	Construction (Design and Management) Regulations
CIRIA	-	Construction Industry Research and Information Association
COPA	-	Control of Pollution Act
COSHH	-	Control of Substances Hazardous to Health
EA	-	Environmental Assessment
EC	-	European Community
EQS	-	Environmental Quality Standard
DoE	-	Department of the Environment
GDO	-	General Development Order
HMIP	-	Her Majesty's Inspectorate of Pollution
HSE	-	Health and Safety Executive
ICRCL	-	Interdepartmental Committee on the Redevelopment of Contaminated Land
HSWA	-	Health and Safety at Work etc. Act
MAFF	-	Ministry of Agriculture, Fisheries and Food
MEWAM	-	Methods for the Examination of Waters and Associated Materials
NAMAS	-	National Measurement Accreditation Service
NAWRO	-	National Association of Waste Regulation Officers
NRA	-	National Rivers Authority
PAH	-	Polyaromatic hydrocarbon
PCB	-	Polychlorinated biphenyl
PDO	-	Potentially Damaging Operation
PTE	-	Potentially Toxic Element
RPB	-	River Purification Board
SAC	-	Special Area of Conservation
SEPA	-	Scottish Environmental Protection Agency
SI	-	Statutory Instrument
SNH	-	Scottish Natural Heritage
SPA	-	Special Protection Area
SSSI	-	Site of Special Scientific Interest
USACE	-	United States Army Corps of Engineers
USEPA	-	United States Environmental Protection Agency
WRA	-	Waste Regulation Authority
WAMITAB	-	Waste Management Industry Training and Advisory Board
WMLR	-	Waste Management Licensing Regulations
WMP	-	Waste Management Paper

Part A Guidance

GUIDANCE

Part A - Contents

GUIDANCE

GUIDANCE

1 Introduction

1.1 BACKGROUND

On 1 May 1994 the Waste Management Licensing Regulations introduced under Part II of the Environmental Protection Act 1990 came into force. On 1 April 1995 the Waste Management Licensing (Amendment etc.) Regulations 1995 came into force. Further amendments have been introduced through the Waste Management Licensing (Amendment No. 2) Regulations 1995. For the purposes of this document "Regulations" will be taken to include all of these Regulations. The Regulations have potentially significant implications for many of those involved with the disposal of dredged material to land.

"Waste" legally means any substance which the producer or holder discards or intends or is required to discard. In practice, this means substances that are not commercially useful in their present form. The Environmental Protection Act 1990 applies to "controlled waste", which covers household, industrial and commercial waste. Waste from dredging operations is legally classified as industrial waste, and is thus "controlled waste" regulated by the legislation.

Prior to the introduction of the Waste Management Licensing Regulations and subsequent amendments, much of the disposal of dredged material to land was largely unregulated. Under the Regulations, unless an operator can demonstrate that the material is not a waste (i.e. it has not been discarded and will be used) or that the disposal option meets one of the specified exemptions, a Waste Management Licence will be required. A Waste Management Licence will involve associated site management responsibilities and costs.

These Regulations will mean that many operators (including waterway owners and managers, dredging operators, waste managers, and others) will now have to apply for a licence(s) and/or register an exemption(s).

This document was prepared in response to a need identified by the dredging industry, waterway operators and the key regulators, the National Association of Waste Regulation Officers (NAWRO) representing the Waste Regulation Authorities (WRA) and the National Rivers Authority (NRA), for clear guidance. The information in this document is current as of the end of 1995.

1.2 PURPOSE AND USE

The Waste Management Licensing Regulations are detailed and complex. There are many ambiguities which will need to be tested in the courts. Notwithstanding this, operators will have to carry out their activities in accordance with the new Regulations. A key factor in any operation is cost, and therefore this document aims to assist the dredging industry and its regulators in interpreting the Regulations and applying them to their own activities in a legally correct and economic manner.

The document has been prepared in two parts. Part A comprises the guidance element of the document and Part B provides supporting information in a number of key areas. Parts A, B and Chapter 3, Legislative and Regulatory Aspects, are identified separately.

The guidance chapters of the document (Part A) have been written to provide a concise explanation of the decision-making process. The flowcharts, checklists, etc. are designed to provide outline guidance on certain aspects of the decision-making process, paying particular attention to the selection of a disposal, use or re-use option, and the planning associated with disposal. As operators become increasingly familiar with the Waste Management Licensing Regulations, the procedures will become more predictable, and it is envisaged that the flowcharts, etc. will be used more as checklists than as guidance.

Part B, the information chapters, will not be relevant to all disposal options, but are intended to assist operators who have a particular question or problem.

1.3   SCOPE OF THE DOCUMENT

1.3.1   Introduction

This document has been prepared to address the disposal to land of material arising from dredging for navigation or flood defence. However, most of the guidance applies equally well to material arising from dredging undertaken for environmental quality improvements or other reasons.

It is recognised that, in some cases, dredging is undertaken to gain construction materials which are not waste. However, the Waste Management Licensing Regulations treat most dredged materials as waste even when they can be disposed of beneficially. The document reflects this.

The authors of this report acknowledge that much of the dredged material being disposed of to land will be characterised as uncontaminated. In many cases, the disposal process for uncontaminated materials will be more straightforward than that for contaminated materials. However, in response to the requirements for environmental protection, safe disposal, etc., in places this document concentrates on guidance relating to the disposal of contaminated materials.

Finally, it should be noted that the report focuses on the requirements of the disposal process which have been introduced or changed in some way by the Waste Management Licensing Regulations. For example, obtaining a new licensed site, or disposal involving a change of use, may require planning permission and the Waste Management Licensing Regulations may mean that new licensed sites are required and, therefore, that more planning applications have to be made. However, the process of applying for planning permission itself has not been changed by the Regulations. Therefore, while it is recognised that the Town and Country Planning process is critical to the disposal option, it is not the purpose of this guidance document to describe the planning process in great detail.

1.3.2   Geographical scope

The Regulations, and hence the information contained in this report, apply to England, Scotland and Wales. Waste Management Licensing is currently administered by WRAs. In the non-metropolitan counties of England, this function is performed by county councils. There are separate WRAs for Greater London, Greater Manchester and Merseyside. In Wales and the other metropolitan counties of England, the WRAs are administered by the district and/or metropolitan councils, and in Scotland they are administered by the islands or district councils (see Section 3.1.1).

The NRA, another key regulator with responsibilities related to the disposal of dredgings, operates in England and Wales, while the Scottish agencies responsible for ground and surface water quality are the River Purification Boards (RPBs). The national nature conservation agencies are English Nature, Scottish Natural Heritage (SNH), and the Countryside Council for Wales (CCW).

From 1 April 1996 the responsibilities of the WRAs, the NRA and Her Majesty's Inspectorate of Pollution (HMIP) in England and Wales were taken over by the Environment Agency. In Scotland their equivalent functions (including those of the RPBs) were taken over by the Scottish Environmental Protection Agency and in Northern Ireland by the Environment and Heritage Service.

Throughout this report, wherever reference is made to the NRA, this should be interpreted as covering the RPB; where reference is made to English Nature this should be interpreted as covering SNH and CCW.

GUIDANCE

1.3.3 Source of dredged material

The Regulations deal with the disposal of dredged material to land (i.e. above the level of mean low water spring tide; see Section 3.17). Disposal to water inland of the tidal limit is treated as "land" for the purposes of the Regulations. This report therefore applies to all dredged material disposed of in these circumstances, irrespective of where the material was dredged from.

1.4 KEY POINTS OF THE WASTE MANAGEMENT LICENSING REGULATIONS

In summary, the key points of the Waste Management Licensing Regulations are:

- a new legal definition of "waste" based on the EC Directive on Waste
- Waste Management Licences are now required for treating, keeping and disposing of waste as well as depositing waste
- a list of exempt activities which do not need a Waste Management Licence, including;
  - spreading dredgings on agricultural land
  - using dredgings for land reclamation
  - depositing dredgings on banks and towpaths
  some of which are new
- a new requirement to register exemptions with the WRA
- applicants for Waste Management Licences must now be fit and proper persons
- Waste Management Licences must protect groundwater from pollution, prevent pollution of the environment, prevent harm to health and detriment to the amenity of the locality
- all holders of waste are subject to a Duty of Care
- new application fees and subsistence charges introduced for Waste Management Licences
- details of Waste Management Licensing to be recorded in public registers
- existing waste disposal licences granted under the old system are converted into new Waste Management Licences
- a requirement for a Certificate of Completion for a licensed facility when the site is no longer needed.

Overall, therefore, the Waste Management Licensing Regulations both introduce new requirements and potentially make the process of disposing dredged material to land more complex. Some existing disposal operations will be affected more than others in terms of the extent to which changes from "current practice" will be necessary. It is likely that there will be an emphasis, at least in the early stages of applying the Regulations, on examining non-waste uses or exemptions from licensing. There will also be an increased incentive to investigate solutions, for example treatment or novel disposal options, which may not previously have been considered, in order to reduce total disposal costs.

GUIDANCE

2 Selection, planning and management of disposal

2.1 THE DECISION-MAKING PROCESS FOR OPTION SELECTION

The overall objective of assessing the various options available for use, re-use, storage or disposal of dredged material is to identify an option which is technically sound, environmentally acceptable, meets relevant health and safety and other regulatory criteria and is financially achievable.

For the purpose of this document:

- "use" is defined as applying to non-waste materials (i.e. those materials which, because they are to be used "beneficially", do not fall under the Regulations at all; see Section 3.2)
- "re-use" applies to those materials which have been treated, stored, etc. under the terms of the Regulations but are subsequently usefully used (i.e. no licence is required for their eventual placement)
- "disposal" refers to all other options; disposal options may be exempt from licensing under the Regulations or a licence may be required.

Figure 2.1.1 illustrates the main steps in the decision-making process required to achieve the above objective. These are:

- overview of project
- information needs: dredged material
- initial review of options
- information needs: receiving site
- environmental appraisal
- option selection
- disposal.

The figure makes reference to the relevant steps in the process covered in Part A of this document. The various interactions necessary as part of this process are shown in Figure 2.1.2.

The following sections develop specific guidance on each of these steps. Relevant flowcharts, checklists, etc. are indicated on Figure 2.1.2 by reference to other figure numbers, box numbers, etc.

2.2 OVERVIEW OF PROJECT

The first step in the decision-making process when considering use, re-use, or disposal options is an overview of the project, identifying the key issues which will have to be resolved in order to satisfy the WRA (and/or other agencies). This also enables resources (e.g. expenditure) to be carefully targeted, preventing possible wasted time and/or money. The applicant can therefore ensure that he/she gains a licence, has an exemption registered or demonstrates a use in the most cost-effective manner.

The identification of issues should include preliminary discussions about the proposals with the WRA and, if a potential receiving site has been identified, possibly other agencies (e.g. the NRA, planners, landowners, etc.). Such discussions (see Section 2.3) will help to ensure that unexpected questions do not arise at a late stage in the proceedings, increasing costs (e.g. in terms of time lost or resources wasted due to reduced economies of scale).

This identification of key issues should also aim to provide initial answers to a number of important questions, examples of which are listed in Box 2.2.1.

Box 2.2.1   Examples of questions to be considered in identifying key issues

---

Quantity and Quality

- approximately what quantity of material is to be dredged?
- what dredging methods might be used?
- is the material fine or coarse?
- is there any reason to suspect that the material might be contaminated?
- have previous analyses been carried out?
- what is the water content of dredgings at time of removal?

Options

- might there be a demand for the material (e.g. as fill material) to keep it within the chain of utility?
- is there a possible exempt use?
- is there an obvious site for disposal?

Background

- what is the distance of the proposed disposal option(s) from the dredging site?
- what information is already available on the dredged material and/or the proposed receiving site(s)?
- what, if any, are the key environmental concerns likely to be?
- are there services on, over or under the dredging or proposed disposal sites?
- are the dredging and proposed disposal sites easily accessible if transportation of the material is required?
- are there any obvious constraints on particular disposal options (e.g. legal, health and safety, etc.)?

Decision-Making

- what alternative disposal options are potentially available?
- who will need to be consulted?
- what is the most likely option for use or disposal of the material?

2.3   CONSULTATION

2.3.1   Introduction

Consultation will play an important role in obtaining a licence, registering an exemption or, possibly, confirming whether or not a material is waste. The operator will need to be sure he/she is operating in the most cost-effective manner within the law. The regulator will need to understand the requirements/objectives and concerns of the operator in order to be able to provide the correct advice. Both will need to understand the requirements (or concerns) of other interested parties.

Consultation will be critical to many of the steps in the decision-making process and detailed requirements are therefore discussed further, as appropriate, throughout Part B.

2.3.2   Consultation by operators

Throughout the preparation of this guidance document, the WRAs have stressed the need for the operator to consult them on his or her (outline) proposals/intentions in respect of a wide range of considerations related to the Waste Management Licensing Regulations. NAWRO is particularly concerned about the complexity of the exemptions to the Regulations. One example of this complexity is that, although both the nature of dredged material and the proposed disposal option may satisfy exemption requirements, the use of lagoons for treatment or storage as part of the proposed disposal option may bring the activity under the licensing system. Further, as discussed in detail in Chapter 3, there are many areas in which the law has yet to be tested. Consultation is important for a successful disposal operation. Early consultation with all interested parties will avoid delay and wasted establishment costs.

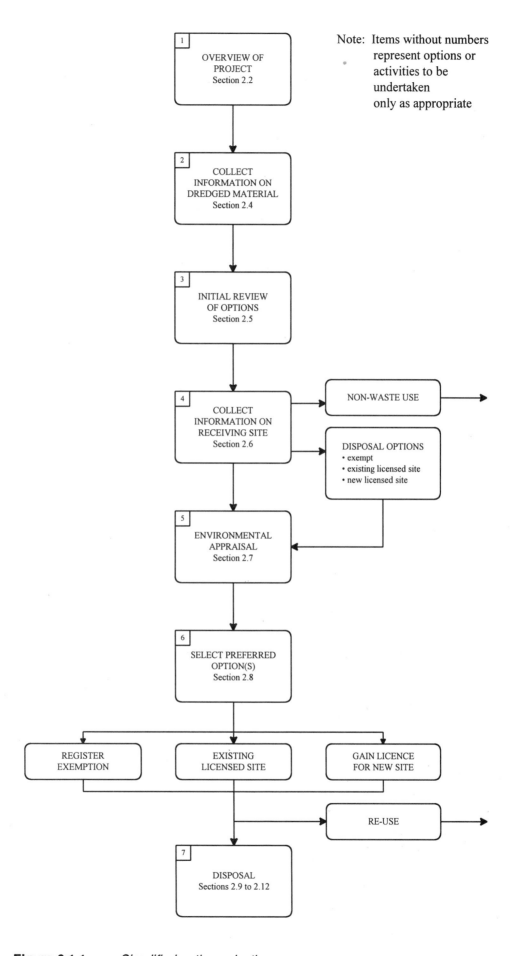

Note: Items without numbers
represent options or
activities to be
undertaken
only as appropriate

Figure 2.1.1   *Simplified option selection process*

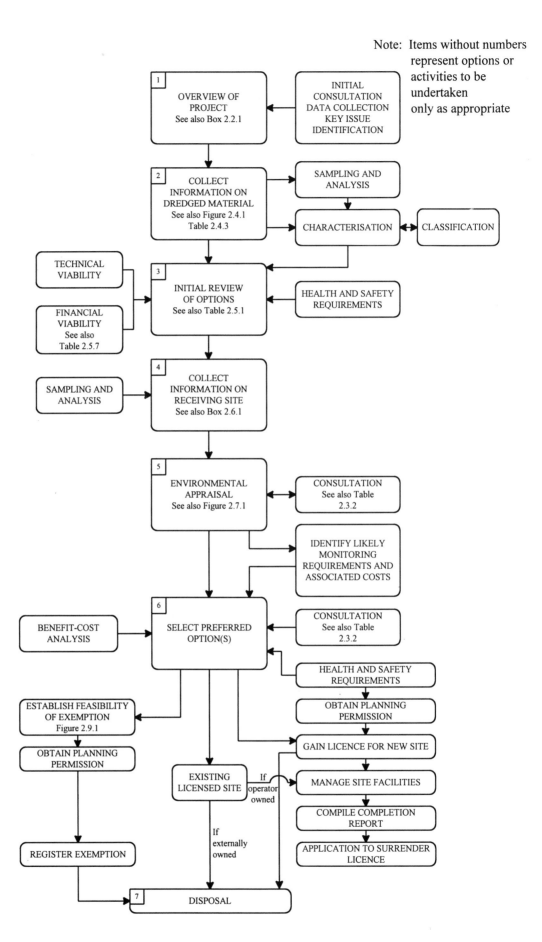

Note: Items without numbers
represent options or
activities to be
undertaken
only as appropriate

Figure 2.1.2   *Option selection process*

Table 2.3.2 identifies potentially important consultees (not in priority order), to be consulted as appropriate, and their role. Other organisations or individuals may also need to be consulted depending on the nature of the site. The planning authority will need to be consulted if planning permission is to be sought (see, for example, Section 3.9.4).

Table 2.3.2   Checklist of consultees

Consultee	Role
WRA*	• statutory body responsible for issuing licences (and registration of exemptions) for all waste management activities including the disposal, handling, treatment and storage of waste. This includes a duty to consult. • consultation of other authorities.
NRA*/River Purification Board (Scotland)**	• statutory consultee for disposal activities requiring a licence • consulted by WRA regarding any waste (disposal) activity, in particular with regard to the risk of contamination of surface waters or groundwaters • issue discharge consents from disposal sites or dewatering areas • statutory consultee for any formal Environmental Assessment work.
Health and Safety Executive	• responsible for implementing Health and Safety at Work etc. Act 1974 (HSWA) and Control of Substances Hazardous to Health (COSHH).
English Nature/Scottish Natural Heritage/Countryside Council for Wales	• statutory consultee when disposal operation may affect a designated conservation site (e.g. Site of Special Scientific Interest) or protected species under 1981 Wildlife and Countryside Act, EC Habitats Directive, etc. Statutory consultee for any Environmental Assessment work.
Local planning authority	• planning permission • conservation sites • landscape issues, recreation issues • tree preservation orders.
English Heritage/Historic Scotland/CADW	• statutory consultee on heritage issues.
Other interested parties	• e.g. owner or designator of non-statutory conservation sites.
Landowner(s)	• rights as landowner.
Consultants	
Environmental	• properly qualified advice • Environmental Assessment/studies (see Section 2.7).
Agricultural	• properly qualified advice on agricultural benefit which may be required by WRA for exemption registration • impacts on agricultural land.
Ecological or local wildlife trust	• properly qualified advice on ecological improvement • impacts on ecological interests.

* As of 1 April 1996 the Environment Agency
** As of 1 April 1996 the Scottish Environmental Protection Agency

2.3.3 Consultation by WRAs

Whilst there is no statutory obligation to consult other regulators when evaluating a notification of an <u>exemption registration</u>, the WRA may choose to consult the NRA where significant water quality issues are involved or other bodies such as English Nature if an exempt activity affects a Site of Special Scientific Interest (SSSI).

GUIDANCE

When evaluating a <u>Waste Management Licence</u> there is a statutory requirement for the WRA to consult the NRA. The Department of the Environment's (DoE) Waste Management Licensing (Fees and Charges) Scheme 1995 (DoE, 1995c) incorporates the costs of involving the NRA in licensing issues. Consultation with bodies other than the NRA will be carried out by the WRA, not the operator, although operators may consult directly. The potential benefits of consultation by the operator (i.e. in ensuring that the necessary information is collected first time and that unnecessary objections are avoided) should be borne in mind.

2.4 INFORMATION NEEDS: DREDGED MATERIAL

2.4.1 Introduction

In order both to identify a suitable disposal option and to satisfy the WRA and/or NRA, the operator will need to investigate and understand various characteristics of the dredged material. This characterisation process (see Chapter 4) is critical to the decision-making process in many respects, not least in determining whether or not a material is contaminated and hence the likely cost of disposal and/or the need for treatment. Of particular relevance in terms of cost are whether the material can be shown to be inert (in which case there are reduced costs involved in its disposal), and whether or not the material is biodegradable (if it is, expenditure on monitoring, site management, etc. will be more onerous). In association with Figure 2.1.2, Figure 2.4.1 helps to define possible information needs.

2.4.2 Information requirements

The amount of "new" information needed in respect of the material to be dredged will be a function, *inter alia*, of the reliability of existing data (see Section 4.3.4). Information on the quantity to be dredged will also be important when disposal options are being considered. Early discussions with the WRA will help to ensure that all essential data requirements are specified in good time.

The level of information required on the nature of the dredged material will vary according to:

- the concerns of the WRA
- possible disposal options (pending the results of any testing, etc.)
- the sensitivity of the receiving site (once known)
- other factors (e.g. the landowner's requirements).

2.4.3 Characterisation

Characterisation involves developing a broad understanding of the dredged material's physical, chemical and biological properties to help evaluate disposal options. In some cases, characterisation will involve the sampling and analysis of the material to be dredged (see Sections 4.3 and 4.4), while in other cases characterisation based on existing information will suffice (e.g. where the dredged material clearly arises from the uncontaminated surrounding land, and/or where testing has been carried out previously). Sampling and analysis may also be necessary if the receiving site is especially sensitive (see Section 2.6).

Where information on the character of the sediment is required (e.g. to prove an exemption), the type of information needed should be defined, if necessary with the assistance of appropriate experts or consultants. This information can then be collected and presented to the WRA and/or the NRA. As the WRA is the responsible regulatory authority, consultation with the NRA should normally take place through the WRA (e.g. proposals should be agreed by the WRA in consultation with the NRA). Some parameters will be of more concern than others because of site specific influences or because of the proposed use or disposal option. For example, organic matter content should help to establish whether a sediment is biodegradable (see also Section 4.6) and may effect contaminant availability. Typical information requirements in this respect are highlighted in Table 2.4.3 and covered in more detail in Part B.

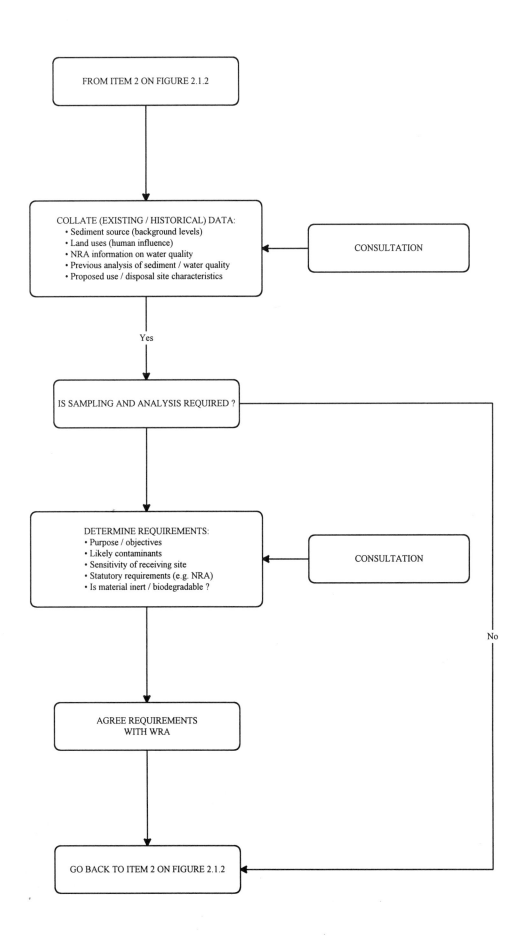

Figure 2.4.1   *Defining information requirements for dredged material*

It is again considered important that, wherever possible, the operator goes to the WRA and/or NRA with suggestions in respect of characterising the material to avoid the cost of re-sampling and re-analysis. The WRA/NRA will assist in determining what parameters should be looked for, but the ultimate responsibility for quantifying and fully describing the waste rests with the producer of the waste. Because a certain parameter has not been analysed for, it does not mean that it is not present in the waste. As indicated above, the use of expert consultants can aid the preparation of a thorough report on the characteristics of the materials. A professional report, whether it is produced by a consultant or operator, will aid the consultation process.

Table 2.4.3   Information requirements for the characterisation of dredged material

Waste characteristic	Parameters of interest						
	Particle size	Contaminants			Nutrients	Organic matter content	Volatile content
		Heavy metals	Organic contamination	Inorganic contamination			
Contamination	✓	✓	✓	✓		✓	
Inert	✓	✓	✓	✓	✓	✓	✓
Biodegradable			?			✓	✓
Benefit to agriculture	✓	✓[1]	?[1]	?[1]	✓	✓	
Ecological improvement	✓	?[1]	?[1]	?[1]	✓	✓	

✓   Probable information requirement
?   Possible information requirement
(1) It may be necessary to prove "no detrimental environmental impacts" as well as demonstrating a "beneficial" use.

2.4.4 Contaminants

It is important to identify the level of contamination in the dredged material as this will help to identify disposal option(s) and hence associated disposal licensing and monitoring costs. The contaminants found in a dredged material will generally reflect the upstream and adjacent uses of the waterway. Relevant background information can therefore help target any sampling and analysis requirements (see Section 4.3.4). The potential risk to the disposal site also needs to be assessed in terms of the contaminants' mobility from the dredged material to the surrounding environment. If appropriate, this assessment could be carried out via leachate testing (see Section 4.4.4 for suggested methods).

2.4.5 Inert waste

Defining a dredged material as inert categorises the waste in a lower cost division under the Waste Management Licensing (Fees and Charges) Scheme 1995 than the cost of non-inert industrial waste. Dredged material has been classified in the Controlled Waste Regulations 1992 as industrial waste.

It is likely that some dredgings will fit the Fees and Charges Scheme definition of inert waste: *"waste which, when disposed of in or on land, does not undergo any significant physical, chemical or biological transformation"*. Proving dredged material to be inert will require detailed consultation with a WRA as well as analytical study and regional variations in interpretation and policy will occur. This is discussed in more detail in Section 4.5.

2.4.6 Non-Inert waste

Non-inert waste is waste which may undergo significant biological, chemical or physical transformation. Biodegradable waste is waste which undergoes significant biological transformation. Therefore, material that only undergoes significant chemical and/or physical transformation may not be inert but at the same time may not be biodegradable (e.g. phosphorus-contaminated waste). Thus non-inert waste may or may not be biodegradable.

2.4.7 Biodegradable waste

Biodegradable waste will normally be subject to landfill gas monitoring and other safety measures if deposited in a licensed site. A biodegradable material contains organic matter which has not been fully decomposed, and has the potential through microbial action to utilise either atmospheric/dissolved (i.e. free) oxygen or any other source of oxygen while the organic matter is being decomposed. There is likely to be a little organic matter in most sediments, and there is some justification for discounting very low quantities of organic matter as constituting a significant biodegradable waste (i.e. indicating no significant "biological transformation"). In this context there is also a criterion that defines non-significant methane production potential. This is less than 10% volatile solids (Waste Management Paper No. 26A; DoE, 1994c). This may not, however, be adequate to define biodegradable without the need for further investigation to prove it, which should involve consultation with the WRA. It is also worth noting that Waste Management Paper No. 27 (DoE, 1994d) states that "*a gas monitoring programme should be incorporated into the design of all operating and proposed landfills irrespective of waste type*". These issues are discussed in more detail in Section 4.6.

2.4.8 Classification

Finally, once any analysis has been carried out, it may be appropriate to compare the results to a classification system to determine whether, according to the system, the material is uncontaminated, contaminated or highly contaminated (see Section 4.7). Such systems should, however, only be used to provide guidance on contamination levels/disposal options and should not be considered to represent rigid and strict definitions (see Section 4.7.6).

2.5 INITIAL REVIEW OF OPTIONS

2.5.1 Introduction

Once information has been collated on the characteristics of the dredged material, an initial review of possible use, re-use, or disposal options should be undertaken. Non-waste (i.e. beneficial) uses (see Section 2.5.2 and 3.2), which are not subject to the Waste Management Licensing Regulations, may represent a potentially viable option for some operators. The various options in respect of waste re-use or disposal, including any prior storage and/or treatment requirements, are shown as a checklist in Table 2.5.1.

The purpose of the initial review of options is to identify and reject options which are unlikely to be realistic, and in turn to focus resources and investigative effort onto potentially feasible options. Where an option appears to be viable but there are problems (e.g. environmental, health and safety, or engineering), special attention must be paid to cost-effective mitigation measures (see, for example, Section 5.4).

It may be difficult to make judgements because of a lack of sufficiently detailed information or because of contradictory views or opinions. In these cases a decision must be taken on the relative benefits of obtaining the necessary data. The review of options is essentially an iterative process and additional information can therefore be obtained and considered at any time.

Table 2.5.1   Checklist of waste re-use or disposal options

OPTION	CHECKLIST OF EXAMPLES		
Storage[1]	• on land (banks, agricultural land, lagoon, pit) • under water (pits, waterways).		
Treatment[2]	• natural dewatering • mechanical dewatering • gravity bed/hydrocyclone (particle size separation) • concentration/destruction techniques (contaminant removal).		
Disposal	Exemption[3]		Re-use[5]
	• bank disposal • agricultural land • land reclamation/improvement. • composting biodegradable waste • use of waste from excavation for construction work		• road construction • landscape/reclamation • use in industries • beach nourishment • river enhancement.
	Licensed site[4] • lagoon • existing landfill site • new landfill site • sea disposal (regulated by MAFF)		

Notes:

(1) see Section 6.2
(2) see Sections 2.5.4 and 6.4
(3) see Sections 3.3 to 3.8 and 6.6
(4) see Section 6.7
(5) see Section 6.3.

It may be particularly important if the registration of an exemption is being considered, or if the material is to be "used" but there is uncertainty as to whether or not such an option will "qualify" as not waste. Consultation with the WRA, NRA, English Nature, etc. (see Section 2.3) is therefore critical, but an independent overview (e.g. by consultants) may be of equal value. In many cases, registering an exemption is likely to be straightforward. In others, however, a great deal of information may be needed (see, for example, Sections 5.5 and 5.6). The cost in time and money of registering an exemption needs to be considered against the cost of licensed disposal.

Key cost considerations (see Section 2.5.7 and Chapter 8) are:

- is the material waste?
- is there an exempt disposal option?
- should the material be treated?
- can the material be deposited at an existing licensed site?
- is a new licensed site required?

The process of answering these questions is illustrated in Figures 2.5.2 and 2.5.3.

2.5.2 Is the material waste?

Waste is defined in the Waste Management Licensing Regulations 1994 as *"any substance or object in the categories set out (in the relevant Schedule to the Regulations) which the producer or the person in possession of it discards or intends or is required to discard"*. As discussed in detail in Section 3.2, if a material is <u>not</u> defined as waste it is not covered by the Waste Management Licensing Regulations.

The definition of waste is very complex. For example, a material does not necessarily cease to be a waste simply because it can be put to a beneficial use. A substance or object should still be regarded as waste where the purpose of any beneficial use is wholly or mainly to relieve the holder of the burden of disposing of it and the user would be unlikely to seek a substitute for it if it ceased to be available.

In order to establish whether or not a material has a use under the terms of the Waste Management Licensing Regulations or whether a licence is required for part(s) of the process, it would be prudent to hold discussions with the appropriate WRA well in advance of the dredging taking place.

2.5.3 Is there an exempt disposal option?

Sections 3.4 to 3.8 describe the exemptions (i.e. disposal options which do not require a licence) to the Waste Management Licensing Regulations. There are various exemptions in respect of agricultural or ecological benefit, land improvement and bankside disposal. An exempt use has to be registered with the WRA, who may require evidence or justification to support an applicant's notification for registration of an exemption (see Figure 2.5.2). In some cases, a certain amount of effort may be required on the part of the applicant in order to convince the WRA of the validity of his/her argument (see Section 2.4 and Chapter 4). The WRA cannot refuse to register particulars that ostensibly meet the terms of an exemption. The WRA may, however, take action for the unlicensed disposal of waste if it believes that there is no entitlement to the exemption.

The DoE's guidance (Circulars 11/94 and 6/95; DoE, 1994; 1995) on benefit to agriculture or ecological improvement states that *"The phrase should be interpreted according to the type of wastes and category of land involved. On land not used for agriculture, only ecological improvement may be relevant."* The spreading of dredged material would have to be in quantities and at frequencies which "convey positive benefits". Therefore appropriate application rates for the dredged material for each soil and each site need to be determined, along with any associated beneficial and detrimental impacts (see Sections 5.5 and 5.6). It is likely that a WRA will require the applicant to take "properly qualified advice" (i.e. from a recognised expert or specialist) on such benefits before making a decision on registering an exemption. It is the responsibility of the exemption applicant to provide such information.

Planning permission may be required if the exemption relates to land reclamation or a change of use (see Section 3.6).

2.5.4 Should the material be treated?

Treatment may allow beneficial re-use of material. Treatment may deal with contaminants, physical grading, or addition of constituents to give the dredgings added value, e.g. as fertiliser.

Treatment of dredged material to remove the contaminated fraction can be achieved by physical, chemical, thermal or biological means. The process prior to any of these means is normally referred to as pre-treatment or sorting of dredged material. Treating contaminated dredged sediment leaves behind waste by-products which need to be dealt with, whichever treatment technology is used.

GUIDANCE

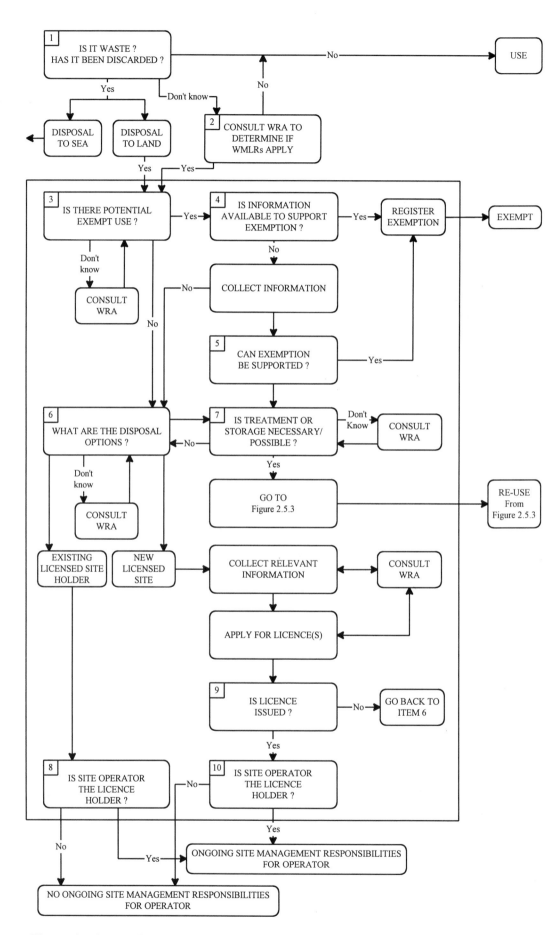

Figure 2.5.2   *Definition of disposal options*

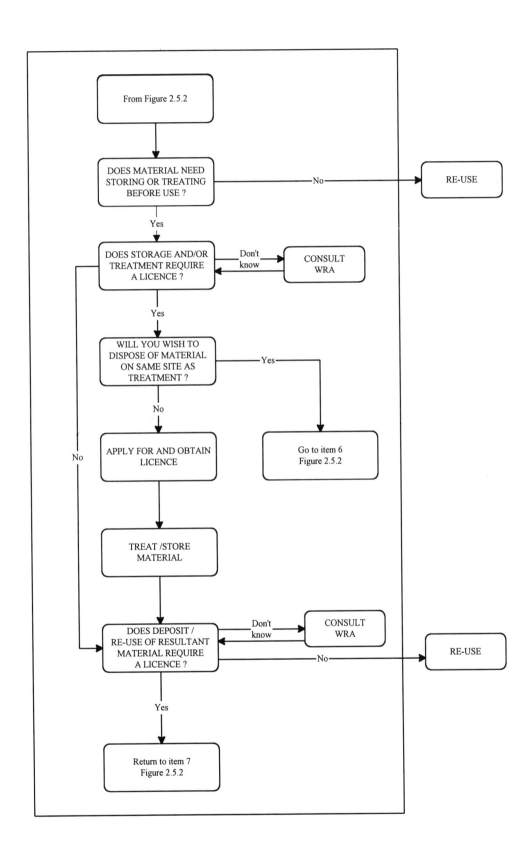

Figure 2.5.3 *Treatment or storage: licensing requirements*

Where a treatment option to decontaminate material is to be considered, the level and type of contaminants will determine the appropriate treatment technology. Treatment of contaminated dredged material may not always be technically or financially feasible. Some of the treatment technologies are regarded as being largely untried (except in pilot studies) and therefore risky. A number of studies have been carried out and reports written on the treatment of contaminated soil and sediment (e.g. Arimshaw *et al.*, 1992; USEPA, 1994; CIRIA, 1995). Deciding on which treatment technology may be applicable for dealing with contaminated dredged sediment will require feasibility studies and possibly pilot scale trials.

As indicated by Figure 2.5.3, re-use options following treatment or storage may nevertheless be worthy of investigation under the Waste Management Licensing Regulations. If the cost of transporting material is high, for example, treatment options or selective dredging techniques might be considered to reduce the amount of material requiring transport to a disposal site. In turn these costs should be compared to the costs of all other disposal options.

If temporary storage and/or treatment is required prior to disposing to an existing licensed site, a Waste Management Licence may be required for these activities.

2.5.5 Can the material be deposited at an existing licensed site?

Material can be deposited at an existing licensed site if the dredged material is acceptable under the licence conditions. Not all disposal sites will be licensed to receive dredgings which may, for example, be considered to be difficult in terms of handling due to their liquid content. The nature of the dredged material (e.g. liquid content, contamination, organic matter, etc.) may affect the costs charged to the disposer by the operator of a licensed site.

Disposal to a licensed site may be a preferred option where:

- the quantity of material is small
- no cost-effective alternative/new disposal site is available
- no exemption is available (i.e. on banks or agricultural site).

Site specific conditions such as prevailing wet weather can affect a site's ability to accept material, irrespective of whether previous arrangements for its "acceptance" have been made.

2.5.6 Is a new licensed site required?

If none of the options outlined in Section 2.5.3 to 2.5.5 is suitable or affordable, an operator may apply for his/her own licensed disposal site. Planning permission may be needed for such a site (see Section 3.9.4). Including the cost of collecting the information needed to satisfy the WRA or Health and Safety Executive (HSE) (see Chapter 7) as part of the application procedure, or obtaining planning permission (if needed), the setting up of a new licensed site can be expensive because of waste management operations such as monitoring and record keeping; fees and charges; and requirements for trained personnel. That said, for a producer of large amounts of dredged material, and/or one who dredges regularly, a new licensed site may indeed be the most cost-effective option.

2.5.7 Cost-effectiveness

The financial implications of each of the options discussed above must be considered. The basic costs, see Table 2.5.7, and any additional costs to meet special technical or environmental mitigation needs, should be estimated in order to arrive at the most cost-effective option. This is discussed further in Chapter 8.

In some cases, the financial assessment may be constrained by a lack of sufficiently detailed information. The cost of information should be set against the possible benefits from it, such as a favourable decision by the WRA.

Table 2.5.7   Cost items associated with different disposal options

Cost	Agricultural land[1]	Reclamation/ Improvement[1]	Bankside disposal[1]	Lagoon[2]	Landfill[2]	Sea disposal[2]	Use/ Re-use[3]
Site Identification/ Feasibility	✓	✓	✓	✓	✓	✓	✓
Sampling/Testing	✓	✓	✓	✓	✓	✓	✓
Agricultural/Other Consultancy	✓	?	?	✓	✓	✓	
Site Design				✓	✓		
Exemption Registration	✓	✓	✓				
Post Exemption Deposit Aftercare	✓	✓	✓				✓
Waste Licence				✓	✓		
NRA Discharge Consent		?		✓	✓		
MAFF Licence						✓	?
Planning Permission		?		✓	✓		
Transport	?	?	?	✓	✓	✓	✓
Transfer	?	?	?	?	?	?	?
Storage Costs	?	?					✓
Treatment	?	?		?	?		?
Spreading Costs	✓	✓	✓				?
Compensation Payments	✓		?				
Land Purchase or Lease	?	?	?	✓	✓		
Site Construction				✓	✓		
Site Operation				✓	✓		
Revenues from Sales		✓					✓
WAMITAB Training				✓	✓		
Site Remediation/ Closure	?	?	?	✓	✓		
Post-closure/ Aftercare				✓	✓		

Key:

✓   indicates that the cost item is likely to be incurred
?   indicates that the cost item may be incurred in some circumstances
(1) exempt
(2) licensed
(3) placement itself outside Waste Management Licensing Regulations although treatment or storage may require a licence.

2.5.8 Need for detailed evaluation

Once the initial review of options has been completed, more detailed investigations may be required before a final choice can be made. In particular, information will be required on the receiving site (Section 2.6) and/or investigations into the possible environmental effects of disposal (Section 2.7). This latter case is particularly important if mitigation measures need to be identified to satisfy the WRA or NRA concerns.

2.6   INFORMATION NEEDS: RECEIVING SITE

Dredged material can potentially interact with the receiving site to cause changes to, or impacts on, the existing environment of or around that site. In order to assess the likely viability of depositing dredged material at a receiving site, various characteristics of that site must be defined. The checklist in Box 2.6.1 identifies some of the key characteristics of the receiving site for which information may be needed, and this is further discussed in Section 5.3. The level of detail required will depend on the requirements of the WRA (and/or NRA); the likely sensitivity of the site and its environs; and/or planning requirements.

Box 2.6.1   Possible information needs to evaluate environmental impacts at a receiving site

Physical characteristics	Land-use
Soil structure/permeability Geology Topography Hydrology Wind direction Rainfall	Land-use at receiving site Services on receiving site Use of adjacent areas Nearest residential area(s) Recreation or amenity uses of receiving or adjacent sites
Water	**Natural environment**
Proximity of surface or groundwater resources NRA groundwater vulnerability classification Quality of water Use of water	Designated nature conservation sites Protected species Protected landscape areas

2.7   ENVIRONMENTAL APPRAISAL

2.7.1 Introduction

The "disposal" of dredged material will usually cause changes to environmental characteristics of the sediment or the receiving site. Such changes may be beneficial (e.g. agricultural or ecological improvements under exemptions; see Sections 5.5 and 5.6) or they may be adverse. Of particular concern to the regulatory authorities is the disposal of contaminated dredged material and the possible consequences for the quality of surface water and groundwater (see Sections 5.7.3 and 5.7.4). The potential for other impacts of both construction and operation (e.g. on the local community, wildlife or landscape) will also need to be considered, although in some cases they will be insignificant.

2.7.2 Environmental appraisal

The process of identifying, evaluating and resolving environmental issues can be referred to as environmental appraisal. Where there is a formal requirement for this, such as may occur when applying for a new disposal site in a sensitive area (see Section 5.2.2), this is known as Environmental Assessment (see Section 3.9.5). Environmental appraisal is discussed in detail in Chapter 5 and a simplified version is shown in Figure 2.7.1.

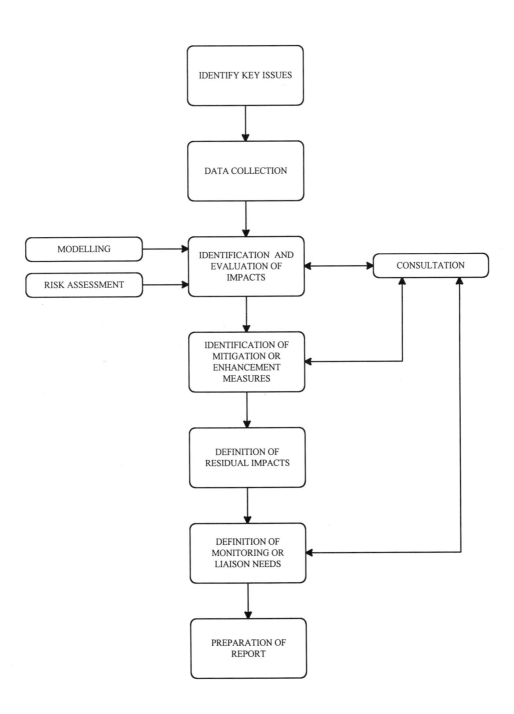

Figure 2.7.1   *Simplified environmental appraisal process*

2.7.3 Scoping and impact evaluation

The first step in the environmental appraisal process is the identification of key issues. As with the overall evaluation process, this is undertaken to ensure that the environmental investigations are as cost-effective as possible. Consultation with interested parties (see Sections 2.3 and 5.1.3) plays a vital role not only in the scoping exercise and the collection of data, but throughout the environmental appraisal process. Once data have been collected (see Sections 2.4, 2.6 and 5.3.2), potential impacts will need to be identified and evaluated. In many cases this will simply involve discussions with relevant organisations, while in others, modelling and/or risk assessment may be necessary (see Section 5.3.3).

2.7.4 Mitigation

Mitigation measures must be investigated where potentially adverse impacts have been identified (see Section 5.4). This is perhaps the most critical part of the environmental appraisal process. The primary mitigation measure is to change the operation to a different location or to a different method to avoid identified impacts.

The careful identification and (planned) implementation of mitigation measures can make the difference between the success or failure of a licence application.

2.7.5 Residual impacts

Once mitigation requirements have been defined and any opportunities for environmental improvement identified, residual impacts of the disposal (i.e. those which will remain after the proposed mitigation measures are implemented) must be highlighted. This is essential in facilitating the decision-making process.

2.7.6 Monitoring and environmental statement

In some cases, the Waste Management Licensing Regulations will set out monitoring requirements (see Section 2.10). In others, monitoring needs will be highlighted by the environmental appraisal process. In all cases, monitoring and liaison needs can usefully be set out alongside a description of the investigations undertaken in an environmental report.

If a formal Environmental Assessment is required (e.g. by the planning authority), this procedure will result in the production of a report known as an Environmental Statement.

2.7.7 Typical impacts

Table 2.7.7 highlights some of the most common environmental impacts of dredged material disposal, together with possible mitigation measures. Environmental impacts are, however, site specific (see Chapter 5). Thus the table is not definitive.

2.7.8 Beneficial impacts

In addition to the above, dredged material may be exempted from the licensing procedure if beneficial impacts to agriculture or ecology can be demonstrated. This is further discussed in Section 2.5.3 and in more detail in Sections 3.5 to 3.7, 5.5 and 5.6.

Table 2.7.7   Typical environmental impacts

Environmental parameter	Impact	Possible mitigation options
Water quality	Contamination of ground or surface water with possible adverse effects on human health, animals and plants	Design modifications; leachate control; lining or capping
Local community	Windblown dust Noise during construction Odour Traffic disturbance/air quality deterioration Loss of access/recreation areas	Careful siting; capping Restricted working hours Gas extraction system Use bulk haul transfer Re-direct footpath; provide alternative recreation areas
Landscape	Visual intrusion	Landscaping Screening Reduction in height/area
Nature conservation	Habitat destruction Species loss	Selective clearance Habitat creation

2.8 OPTION SELECTION

2.8.1 Introduction

Once the necessary level of information has been collated, an informed review of potentially viable use, re-use or disposal options as outlined in Table 2.5.1 can take place so that a preferred disposal route (see Figure 2.5.2) may be chosen. This review involves four components: the assessment of technical criteria, environmental acceptability, cost-effectiveness, and consultation.

2.8.2 Technical criteria

In the first instance, the technical (or engineering) viability of the different options must be confirmed. Any special measures necessary to meet the Waste Management Licensing Regulations and/or health and safety objectives (see Chapter 7) must be defined. This might include, for example, the use of a qualified engineer to design retaining bunds, the establishment of site safety rules, the provision of first aid facilities, etc. Consultation with the HSE, either directly or through the WRA, may be appropriate at this stage.

2.8.3 Environmental acceptability

Secondly, each option must be assessed in terms of its environmental acceptability and in terms of any mitigation measures likely to be required. Of particular importance in this respect will be the interaction between the dredged material and the receiving environment. The actual effect (if any) on the receiving site will consist of physical, chemical and/or biological changes following deposition of the dredged material.

2.8.4 Cost-effectiveness

As described in Section 2.5.7, cost is an essential component in the decision-making process. The increased complexity introduced by the Waste Management Licensing Regulations means that, in some cases, operators will incur costs for new activities or analyses and some elements of previous disposal practice may no longer be acceptable. It may therefore be worthwhile in some cases to investigate treatment or novel disposal options (see Chapter 6). The costs of all disposal options should then be compared and an affordable option found which meets the necessary requirements (see also Chapter 8).

GUIDANCE

2.8.5 Consultation

Prior to embarking upon this procedure the regulatory authorities stress the need for organisations to consult them as early as possible on the planning of disposal of dredged material. The local WRA will always endeavour to deal with enquiries promptly but sometimes a heavy workload may cause delay. If this should happen, it is best to identify the key issues and consult directly with the NRA or other statutory bodies.

Written evidence of any consultation carried out during option selection should be submitted with licence application or exemption registration details. This should help to ease the WRA assessment of the application or registration by demonstrating the reasoning behind, and justification for, the option selected.

2.9 APPLICATION AND NOTIFICATION

2.9.1 Introduction

The notification procedures required to register an exemption under the Waste Management Licensing Regulations for the treatment, storage or disposal of waste are illustrated in Figure 2.9.1.

2.9.2 How an exemption is registered

The organisation intending to carry out an exempt activity is required to register with the local WRA (see Section 3.4). The notification requirements for each exemption option are shown in Figure 2.9.1. The WRA is particularly concerned about the timing and the duration of the exempt activity. The WRA may request additional information, for example, analytical data and details of agricultural benefit for spreading on agricultural land.

2.9.3 Is planning permission needed for an exemption?

Planning permission is only required if there is a change of use. Therefore, of the exempt activities listed in Figure 2.9.1, only land reclamation may require planning permission to be obtained before registering an exemption (see Section 3.6). Certain authorities (such as BW and the NRA) may have powers under the planning legislation to proceed without permission, but it would be sensible to notify the planning authority of the proposed work and obtain confirmation that it is agreed permitted development.

2.9.4 How to apply for a Waste Management Licence

The formal application and notification procedures required to obtain a licence under the Waste Management Licensing Regulations are illustrated in Figure 2.9.4.

The person intending to occupy the land for which a Waste Management Licence is required is responsible for obtaining it (see Section 3.9). The WRA would usually expect initial discussions to occur several months before a written application is made. During this time proposals should be put to the WRA and guidance will be given on the information required to support them. Formal applications would normally include the following:

- type of waste and quantities
- location and boundaries
- land ownership
- planning permission or other certificate of lawful use.
- information on convictions, technical competence and financial provision
- physical site assessment
- working plan containing engineering details of the site, how the activities will be carried out and the environmental monitoring regime.

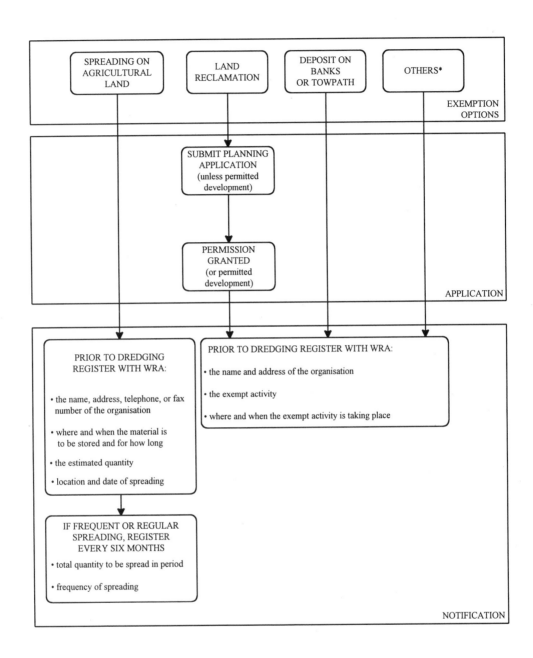

EXEMPTION OPTIONS
- SPREADING ON AGRICULTURAL LAND
- LAND RECLAMATION
- DEPOSIT ON BANKS OR TOWPATH
- OTHERS*

APPLICATION
- SUBMIT PLANNING APPLICATION (unless permitted development)
- PERMISSION GRANTED (or permitted development)

NOTIFICATION

PRIOR TO DREDGING REGISTER WITH WRA:

- the name, address, telephone, or fax number of the organisation
- where and when the material is to be stored and for how long
- the estimated quantity
- location and date of spreading

PRIOR TO DREDGING REGISTER WITH WRA:

- the name and address of the organisation
- the exempt activity
- where and when the exempt activity is taking place

IF FREQUENT OR REGULAR SPREADING, REGISTER EVERY SIX MONTHS
- total quantity to be spread in period
- frequency of spreading

Key:

* Other exemptions: Spreading on operational land
(see Section 3.3)   Composting
                        Storage and use of waste from excavations for consruction work
                        Treatment of waste plant matter

Figure 2.9.1   *Notifications required for exemption options*

Planning permission (see Section 3.9.4) is often sought in parallel with applying for a Waste Management Licence. A large amount of information is required to support both planning and Waste Management Licence applications and therefore it is often more practical and cost-effective to manage these aspects concurrently.

A Waste Management Licence will be issued only if planning permission has been granted by the planning authority or there are permitted development rights under a General Development Order (GDO), and only to a person able to prove occupancy of the land. It is common practice to have an agreement whereby occupancy of the land is given if the application for a Waste Management Licence is successful.

The WRA must notify an applicant of its decision within four months of receiving the licence application, or seek an extension of time with the applicant. Providing the information requested has been satisfactorily supplied, the only grounds on which a licence may be refused are pollution of the environment, harm to human health, or serious detriment to the amenities of the locality (unless the activity is authorised by planning permission).

Details of the licensing of waste management facilities are given in the DoE publication Waste Management Paper No. 4 (DoE, 1994b).

2.9.5 Who can hold a Waste Management Licence?

A WRA must satisfy themselves that an applicant for a Waste Management Licence is a "fit and proper person" to hold it (see Section 3.10). An applicant should not have been convicted of a serious offence under environmental legislation, he/she must be technically competent (normally requiring a Waste Management Industry Training and Advisory Board (WAMITAB) Certificate) and adequate financial provision must be made to meet the obligations of the licence.

2.9.6 How much information is needed?

The nature of the dredgings (Section 2.4) and of the site requiring a licence (Section 2.6) will determine how much detail will be required. However, some data must be obtained before the WRA can issue a licence, for instance, if planning permission is required. The time required for the planning, preparation and submission of a planning application should not be underestimated. The planning authority may require significant information on which to make a decision, including an assessment of the potential environmental impacts of the proposed disposal option. Seeking planning permission is therefore often the cause of real uncertainty and substantial costs.

2.9.7 Working plan

The working plan is the site operator's document. It shows how the licence applicant proposes to prepare, develop, operate, monitor, restore and eventually complete the site facility, whether it is a permanent disposal site or a waste transfer site. Two types of information are contained in it:

- detailed drawings of the engineering of the site (e.g. access roads, fencing, structures, embankments, levels, soil conditions, etc.)
- detailed descriptions of the way the activities on the site are to be carried out (hours of opening, amounts of waste, record keeping, etc.).

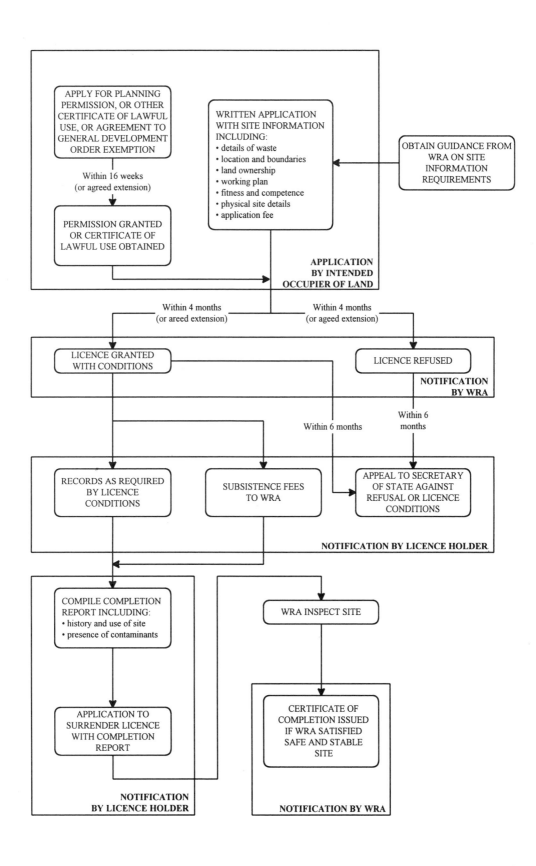

APPLY FOR PLANNING
PERMISSION, OR OTHER
CERTIFICATE OF LAWFUL
USE, OR AGREEMENT TO
GENERAL DEVELOPMENT
ORDER EXEMPTION

WRITTEN APPLICATION
WITH SITE INFORMATION
INCLUDING:
• details of waste
• location and boundaries
• land ownership
• working plan
• fitness and competence
• physical site details
• application fee

OBTAIN GUIDANCE FROM
WRA ON SITE
INFORMATION
REQUIREMENTS

Within 16 weeks
(or agreed extension)

PERMISSION GRANTED
OR CERTIFICATE OF
LAWFUL USE OBTAINED

**APPLICATION
BY INTENDED
OCCUPIER OF LAND**

Within 4 months
(or areed extension)

Within 4 months
(or ageed extension)

LICENCE GRANTED
WITH CONDITIONS

LICENCE REFUSED

**NOTIFICATION
BY WRA**

Within 6 months

Within 6
months

RECORDS AS REQUIRED
BY LICENCE
CONDITIONS

SUBSISTENCE FEES
TO WRA

APPEAL TO SECRETARY
OF STATE AGAINST
REFUSAL OR LICENCE
CONDITIONS

NOTIFICATION BY LICENCE HOLDER

COMPILE COMPLETION
REPORT INCLUDING:
• history and use of site
• presence of contaminants

WRA INSPECT SITE

APPLICATION TO
SURRENDER LICENCE
WITH COMPLETION
REPORT

CERTIFICATE OF
COMPLETION ISSUED
IF WRA SATISFIED
SAFE AND STABLE
SITE

**NOTIFICATION
BY LICENCE HOLDER**

NOTIFICATION BY WRA

Figure 2.9.4   *Applications and notifications required for a Waste Management
Licence*

2.9.8 The difference between licence conditions and the working plan

The two documents are quite separate. The licence conditions stipulate the standards (set by a WRA) to which the site must be operated, while the working plan states how the operator proposes to meet these standards. In practice the licence conditions are often linked to a specific element of the submitted and agreed working plan. At a large site, or one to be used over a long period of time, the working plan is likely to be revised frequently as the facility evolves.

2.9.9 Action necessary when the site is no longer needed

A Waste Management Licence cannot be surrendered until the WRA has inspected the site and is satisfied that the condition of the land is unlikely to cause pollution of the environment or harm to human health. This may incur significant costs as the WRA will require proof that the land is stable and safe. It will involve the production of a completion report covering the history and use of the site and, most importantly, the presence of contaminants. Therefore, there will be a period of time (which may be significant) between the site operation ceasing and licence surrender. During this period the operator will continue to have responsibility for maintaining the site in accordance with the licence conditions including any monitoring which is required. When the WRA has accepted the surrender of a licence it will issue a Certificate of Completion and the liabilities of the operator under the licence will cease.

2.9.10 Costs of a Waste Management Licence

The application fees and continuing charges for Waste Management Licences are payable to the WRA. They are determined in accordance with the current Fees and Charges Scheme which lists the sums payable in respect to the treatment, keeping and disposal of different types of waste (see also Section 3.13).

2.9.11 What is required under Duty of Care?

Duty of Care is imposed by statute (Environmental Protection Act, 1990) on all holders of waste to take all reasonable steps to prevent unauthorised handling of waste, to prevent its escape and to only transfer waste to an authorised person, such as a Waste Management Licence holder or a registered carrier of waste.

Sufficient written information must be supplied with the waste on transfer to enable the recipient to comply with their own Duty of Care (see Section 3.11).

2.10 MONITORING

2.10.1 Is it necessary to monitor?

If an option for dredged material disposal is selected which involves a Waste Management Licence, conditions will often be attached by the WRA to monitor any possible effects on the environment by leachate and/or methane generation. In the case of leachates, this monitoring is likely to have been requested by another regulatory body, the NRA, which will be concerned about monitoring groundwater quality. An operator may be required by a licence to carry out background monitoring of groundwater and/or landfill gas for a significant period of time prior to the commencement of waste disposal activities.

The requirements of which parameters to test for, and guidance on the frequency of monitoring are set out in the following Waste Management Papers:

- No. 4   *Licensing of waste management facilities*
- No. 26   *Landfilling waste*
- No. 26A   *Landfill completion*
- No. 26B   *Landfill design, construction and operational practice*
- No. 27   *Landfill gas.*

The information given in these papers is complex. Paper 4 for example also requires regular inspections with regard to the stability of a containment. In practice the requirements of the WRA, through the licence conditions, are likely to be specific to the characteristics of waste and the site. Sections 5.4.6 and 5.4.7 discuss monitoring requirements in more detail. The DoE are producing a new waste management paper (No. 26D), currently in draft, on landfill monitoring.

2.11   CONTRACTS AND DUTY OF CARE

It is important for the employer to ensure that the contractor is obliged to conform in all respects with the provisions of any statutes relating to the work, including the Waste Management Licensing Regulations and the Duty of Care requirements. The use of standard conditions of contract, such as the Institute of Civil Engineers 6th Edition or any later update, is recommended. The practice of commissioning (dredging and) disposal operations without including a warranty by the contractor, requiring him/her to conform with current legislation and indemnify the employer against any penalties, should be avoided. It should be noted that both the Duty of Care owned by a producer or holder of waste and the obligation to obtain a Waste Management Licence cannot be avoided by making alternative arrangements in a contract between employer and contractor.

A contractor who carries out a dredging and disposal operation on behalf of another party (i.e. a client) will qualify as a "producer" of waste and will be subject to the Duty of Care. Where the client arranges a second contractor to receive and dispose of the waste that second party will be a broker and therefore have a Duty of Care as well. If the first contractor produces and directly arranges disposal with no reference to the second party then the second party will acquire no Duty of Care under Section 34 of the Environmental Protection Act 1990.

The scope of the Duty of Care has not been fully tested in the courts and landowners should always be aware that waste produced by others on their land may be traced back to them should eventual disposal be found to be inappropriate. Employers and landowners would be prudent to take reasonable steps to ensure that all contractors employed comply with the Duty of Care.

2.12   DUTY OF CARE AND STANDARD PROCEDURES

2.12.1 Records

The Environmental Protection (Duty of Care) Regulations 1991 (SI 1991 No. 2839) (made under Section 34(5) of the Act) require all those subject to the Duty to make records of waste they receive and consign, keep the records and make them available to WRAs (see Section 3.11). The Duty of Care Regulations require these records, such as transfer records or descriptions of waste, to be kept for at least two years.

2.12.2 How do records help control the process?

The documentation allows each handler of the waste to understand the nature and destination of the material and therefore to meet the Duty of Care obligation. Records will also become a source of information should there be a need either to investigate site problems at a later stage or to determine the origin of a particular material (e.g. this may arise out of monitoring and the detection of unforeseen contaminants at a landfill site).

2.12.3 Responsibility for the records

All waste holders have a responsibility for making and retaining records. However, the quality of the records and the success of the whole operation are highly dependent on the waste producer describing the waste and the intended destination fully. Insufficient details may result in a subsequent breach of the Regulations (and later in the disposal process), for which the waste producer may then be held liable.

2.12.4 Standard procedures

Well defined and established sampling and monitoring techniques (Section 5.4.7) are essential to assure the WRA that the licence conditions are being adhered to. The adoption of quality procedures for monitoring and control will also avoid potential problems (such as undetected contaminants) and thereby reduce costs overall.

Records of any monitoring and sampling, including results, should be kept by the waste holder/site operator.

2.12.5 Licence documentation

Copies of the licence documentation should be available on site, with care being taken to keep the original safe. Copies of the different issues of the working plan throughout the life of a licensed site should be kept on file.

2.12.6 Other benefits from keeping documents

Holders of existing Waste Disposal Licences are, by and large, deemed to be "fit and proper persons" until they commit an offence or their responsibilities change. If a licence applicant has been convicted in the past, the WRA will use its discretion to decide whether or not the conviction forms sufficient grounds to refuse a Waste Management Licence. Good documentation of past activities, therefore, should provide evidence to support an application to regain a Waste Management Licence should the need arise.

Part B   Supporting Information

Part B - Contents

3   Legislative and regulatory aspects of dredging disposal

3.1   INTRODUCTION

The Waste Management Licensing Regulations 1994, SI 1994 No. 1056, which became law on 1 May 1994, prescribe the details of the Waste Management Licensing system introduced under Part II of the Environmental Protection Act 1990. The Regulations are applicable within England, Wales and Scotland, and implement the European Community Directive 75/442/EEC on Waste, as amended in 1991 by Directives 91/156/EEC and 91/692/EEC. Amendments to the Regulations have been made by the Waste Management Licensing (Amendment etc.) Regulations 1995, SI 1995 No. 288, and the Waste Management Licensing (Amendment No. 2) Regulations 1995, SI 1995 No. 1950.

3.1.1   Administrative framework

Waste Management Licensing is administered by WRAs. In the non-metropolitan counties of England, this function is performed by county councils. There are separate WRAs for Greater London, Greater Manchester and Merseyside. In the other English metropolitan counties (West Midlands, Tyne and Wear, West Yorkshire and South Yorkshire) and in Wales, the WRAs are administered by the district and metropolitan councils, and in Scotland they are administered by the islands or district councils. In Tyne & Wear, West Yorkshire and South Yorkshire the district councils have delegated responsibility for the waste regulatory function to the Waste Management Joint Committee under the 1972 Local Government Act, Sections 102 and 101. In Tyne & Wear, Gateshead Metropolitan Borough Council, via the Tyne & Wear Waste Regulation Authority, acts as the lead authority on all waste regulation matters; in South Yorkshire the South Yorkshire Waste Regulation Unit (based for administrative reasons at Rotherham Metropolitan Borough Council) acts as agent to the metropolitan district councils; and in West Yorkshire the West Yorkshire Waste Regulation Authority (based at Wakefield Metropolitan Borough Council and administered by the West Yorkshire Waste Management Joint Committee) has responsibility for waste regulation. In the West Midlands, a large part of the waste regulatory function has been delegated by the districts to the Joint Committee, and this part is implemented by the West Midlands Hazardous Waste Unit (based at Walsall Metropolitan Borough Council), but the districts still maintain overall responsibility for waste management.

Under the Environment Act 1995, these responsibilities were transferred on 1 April 1996 to the Environment Agency in England and Wales, to the Scottish Environmental Protection Agency (SEPA) in Scotland and to the Environment and Heritage Service in Northern Ireland. The Environment Agency also replaced the NRA, and SEPA took over the functions of the RPBs. Prior to 1 April 1996, however, the boundaries of WRA jurisdiction were co-extensive with local government areas, and consequently, where land or waterways crossed those boundaries, they were subject to the control of more than one WRA. At present, the boundaries of the waste regulation departments within the Environment Agency are continuing to be managed co-extensively with local government areas within each Agency region. Therefore, cross boundary jurisdiction can still apply between different waste regulation departments and Environment Agency regions.

3.1.2   Official guidance

The Waste Management Licensing Regulations are supplemented by other statutory and non-statutory documents. The Waste Management Licensing (Fees and Charges) Scheme 1995, which prescribes the application fees and subsistence charges for licences, is made by the Secretary of State under powers provided in the Environmental Protection Act 1990, and therefore has the force of law. DoE Circular 11/94 provides explanatory guidance on the effect and interpretation of the Waste Management Licensing Regulations. It is directed primarily towards WRAs, and contains the views of central government on how those bodies should exercise their powers. Although the Circular is not a legal document, it is likely to be highly persuasive in practice and reflected in the decisions of the Secretary of State on appeals. Nevertheless, it is open to the courts to reject the interpretation of the Circular on points of law.

Further advice about amendments to the Regulations is contained in DoE Circular 6/95. There are also several Waste Management Papers published by the DoE. Waste Management Paper No. 4 provides statutory guidance on the drafting of Waste Management Licences, No. 26A is a technical memorandum on assessing the completion of licensed landfill sites, and No. 26B is a practical guide to the design, construction and operation of landfills. WRAs are required by law to take account of the guidance in these Waste Management Papers.

3.1.3 Statutory Interpretation

The Regulations are detailed and complex. The few provisions dealing specifically with the disposal of dredgings contain undefined terms which will make their interpretation difficult until consistent practices have been adopted by WRAs. In many cases, the WRA (or the Secretary of State on appeal) will decide the scope of these terms. Although the interpretation of a rule of law adopted by a public authority can be challenged before the courts, they will only overrule a decision of the authority if it is clearly wrong in law or is wholly unreasonable. Moreover, the courts will not interfere with findings of fact rather than law.

In order to establish the effect of the Regulations on the treatment, storage and disposal of waterway dredgings, it is best to approach the subject in three stages:

1. It is necessary to determine whether a particular dredged material constitutes "controlled waste". If it does not, the Regulations are inapplicable, and there is no need to consider them further.

2. If the dredgings are controlled waste, they may nevertheless fall inside a category that exempts them from the requirement of a Waste Management Licence, although the exemption must be registered.

3. If the operation is not exempt, it will be necessary to apply for a licence (or dispose of the material at an existing licensed site), and the provisions of the Waste Management Licensing system must be followed.

In practice, it is advisable to consult or request written confirmation of advice from the WRA on the above. Each of these is considered in the following sections of this chapter.

3.2 THE DEFINITION OF "WASTE"

3.2.1 Introduction

The Waste Management Licensing Regulations only apply to substances that are legally regarded as "Directive waste", which is defined in Regulation 1(3). "Directive waste" comprises any substance or object listed in Schedule 4 Part II of the Regulations which the producer or person in possession of it discards, or intends or is required to discard. None of the categories of waste in Schedule 4 Part II specifically refers to dredgings, but they are inevitably covered by the residual category Q16, which applies to any materials, substances or products not contained in the other categories (although certain substances are excluded under Article 2 of the EC Directive on Waste). The applicability of the Regulations to dredgings therefore depends on the circumstances in which they can be described as "discarded".

In contrast, the Environmental Protection Act 1990, Part II, applies to "controlled waste", which is defined in section 75 of that Act as "household, industrial and commercial waste", but excludes waste from mines, quarries and premises used for agriculture. This is different from the definition of "Directive waste" in the Waste Management Licensing Regulations. Although "controlled waste" cannot now include anything that is not "Directive waste", the reverse is not legally true, and so not all "Directive waste" is currently "controlled waste" for the purposes of the Environmental Protection Act. It is important, because the "Duty of Care", discussed in Section 3.11, applies to "controlled waste" rather than "Directive waste". Any "Directive waste" that is not also "controlled waste" will not be subject to the provisions of the Environmental Protection Act until the definition of "controlled waste" in that Act is amended by Parliament.

3.2.2 Official guidance

DoE Circular 11/94 contains guidance on the DoE's interpretation of the concept of "waste", which is based on the objectives of the EC Directive. The Circular concludes that the purpose of the Directive is to treat as waste those substances or objects which fall out of the normal commercial cycle or out of the chain of utility. Accordingly, the following question should be asked:

• has the substance or object been discarded so that it is no longer part of the normal commercial cycle or chain of utility?

It follows that where dredgings are simply abandoned or dumped (e.g. to landfill) either by the producer or another party they will constitute waste. Their status is more complicated, however, where they are to be re-used for other purposes, and may differ according to whether the re-use is by the producer or another party (see Box 3.2.2).

Box 3.2.2   Definition of waste

Although the Department of the Environment's interpretation of the meaning of "waste" can be expected to be followed by WRAs, a legally definitive view can only be taken by the national courts and the European Court of Justice. The European Court of Justice has so far adopted a very broad interpretation of the concept of "waste" under the EC Directive, notably in the case of Vessoso and Zanetti [1990] ECR 1461. Earlier British cases, which concerned the previous waste disposal licensing system under the Control of Pollution Act 1974, concentrated solely on the intention of the producer, and regarded material which the producer intended to discard as waste even if the recipient used it for a positive purpose: Long v Brooke (1980) CrimLR.109; Kent County Council v Queenborough Rolling Mill Co (1990) 89 LGR 306. It would be surprising if the EC Directive were to have a more restrictive meaning than earlier British law, and the possibility remains that the courts could still rule that dredgings for which the producer has no use may become waste irrespective of the intention of the person to whom they are transferred.

3.2.3 Use of dredgings by producer

If dredgings are put to a <u>beneficial use</u> by their producer (e.g. positive engineering purposes on the producer's own land) without the need for further treatment to eliminate or diminish environmental risk (e.g. from contaminants), the Department of the Environment's guidance implies that they should not amount to waste provided that the beneficial use is not merely <u>incidental</u>. A beneficial use may be regarded as incidental if the producer's purpose is wholly or mainly to avoid the burden of disposing of the dredgings and he/she would be unlikely to seek a substitute for the dredged material if it did not exist. Where the reuse is incidental in that sense, the dredgings would instead be classed as waste. However, it is advisable for definitions of beneficial use to be applied only after discussion with the WRA.

3.2.4 Use of dredgings by another party

If the producer of dredgings transfers them to another party who proposes to use them for beneficial purposes (e.g. as raw material for construction or as fertiliser on agricultural land), the Department of the Environment's guidance implies that their status will depend partly on whether they must first be subjected to a <u>specialised recovery operation</u> in order to remove contaminants. Where dredgings can only be re-used after such reprocessing, they will be regarded as waste until that process has occurred. If, on the other hand, they are re-usable without the need for a specialised recovery operation, they may never become waste. However, in this case, both the producer and the recipient must intend that the dredgings be put to a beneficial use. A use is unlikely to be regarded as beneficial if the primary motive of the recipient is to relieve the producer of the burden of disposal and he/she would be unlikely to seek a substitute material if the dredgings were unavailable. Whether or not payment is made by either party will also be relevant, although not conclusive, evidence of intention. Again, discussion and agreement with the WRA on the status of a use are highly advisable.

3.3 EXEMPTIONS FROM LICENSING

Schedule 3 of the Regulations lists activities which are exempted by Regulation 17 from the need for a Waste Management Licence. It does not, however, automatically follow that without an exemption the activity would otherwise require a licence, and therefore no firm conclusion can be drawn from the existence of an exemption that the material involved legally constitutes waste. Three of the exemptions (Schedule 3 paragraphs 7(1), 9 and 25) expressly relate to the disposal of dredgings, and several others are potentially relevant. These are described in Sections 3.5 to 3.8 below and some are discussed further in Sections 5.5 and 5.6. In view of the variability of dredged material, it is advisable to keep the status of any dredging operations under review, and to consult the WRA on whether they are licensable or exempt.

There are two overriding limitations on the availability of any exemption:

1. The majority of exemptions do not apply if the activity involves "special waste" (i.e. specific hazardous wastes currently defined in the Control of Pollution (Special Waste) Regulations 1980, SI 1980 No. 1709). Seven exemptions (Schedule 3 paragraphs 17, 36, 38-39 and 41-43 of the Waste Management Licensing Regulations) include special waste, but none of these relates specifically to dredgings. The definition of special waste may change as a result of a planned revision in the near future.

2. No exemption may be claimed by an establishment or undertaking (i.e. an organisation rather than a private individual) unless the disposal or recovery of the waste involved is consistent with the objectives of the EC Directive as defined in Schedule 4 Part I paragraph 4(1)(a) of the Regulations. These objectives are:

    *   ensuring that waste is recovered or disposed of without endangering human health and without using processes or methods which could harm the environment and in particular without:

        -   risk to water, air, soil, plants or animals
        -   causing nuisance through noise or odours
        -   adversely affecting the countryside or places of special interest.

Thus, even if the express terms of an exemption are satisfied, it is still possible for the benefit to be denied to an organisation in a particular situation where it might otherwise conflict with the policy of the EC Directive, for example:

*   in a sensitive location
*   where the quantity or quality of waste involved might have an unacceptable impact on the environment.

Again, the qualifying criteria on which a judgement is based make discussion with the WRA essential.

3.4 REGISTRATION OF EXEMPTIONS

Since 1 January 1995, establishments or undertakings (i.e. organisations such as authorities, companies and dredging contractors, but not individuals such as private landowners) currently carrying out exempt activities are required by Regulation 18 to register with the WRA. This does not apply to the spreading of dredgings on agricultural land under Schedule 3 paragraph 7 (see Section 3.5), since a separate system of prior notification exists in that case, which achieves the same purpose. Failure to register is a criminal offence subject to a maximum fine £10 in the case of the exemptions relevant to the disposal of dredgings. The maximum fine was reduced from £500 on 1 April 1995 by the Waste Management Licensing (Amendment etc.) Regulations 1995, SI 1995 No. 288. It should be noted that it is the organisation proposing to carry out the exempt activity that must register. This will not necessarily be the owner of the land on which dredgings are spread, but may be a contractor carrying out dredging operations.

A WRA has a statutory duty to enter the relevant particulars if they are notified of it in writing, or if they become aware of it by other means. The register must contain the following:

- the name and address of the establishment or undertaking
- the activity which constitutes the exemption
- the place where the activity is carried on.

WRA cannot refuse to register particulars that meet the terms of an exemption. Nevertheless, if an organisation claims an exemption to which it is not entitled, it runs the risk of prosecution for the unlicensed disposal of waste, and would therefore be well advised to consult the WRA in order to ascertain that body's view of the legal status of its proposals.

3.5 SPREADING ON AGRICULTURAL LAND

3.5.1 Introduction

Schedule 3 paragraph 7(1) of the Regulations exempts the spreading of dredgings from any inland waters on land which is used for agriculture, subject to certain conditions. The exemption is only available if the spreading is carried out by, or with the consent of, the occupier of the agricultural land, or if the party doing it is otherwise entitled to do so on that land (e.g. under statutory powers). An exemption cannot be claimed if adverse environmental consequences will arise from the spreading, as described in Section 3.3.

3.5.2 The meaning of "inland waters"

"Inland waters" are defined in Regulation 1(3) by reference to their meaning in Acts of Parliament concerning water resources, and they have a wider interpretation in England and Wales compared with Scotland. The definitions were not originally devised for the purposes of Waste Management Licensing, and they present problems of interpretation and compatibility when used in a different context. In England and Wales, "inland waters" means:

- any river, stream or other watercourse, whether natural or artificial and whether tidal or not
- any lake or pond, whether natural or artificial, or any reservoir or dock
- any channel, creek, bay, estuary or arm of the sea.

This definition is taken from the Water Resources Act 1991, Section 221(1), and applies not only to most non-tidal and fresh waters, but also to some tidal waters. While estuaries and bays are specifically included, the legislation is imprecise about their seaward extent, but will arguably cover any tidal waters that are semi-enclosed or associated with the land. In each case, the boundaries will depend on the local situation.

In Scotland, the definition of "inland waters" is taken instead from the Control of Pollution Act 1974, section 30(A), where it means:

• the waters of any relevant loch or pond
• so much of any relevant river or watercourse as is above the fresh-water limit.

The fresh-water limits of Scottish rivers and watercourses are depicted on official maps deposited with River Purification Boards by the Secretary of State for Scotland. The term "relevant" applies to all fresh-water rivers and watercourses, including underground or artificial ones, but excluding public sewers (and sewers or drains that drain into public sewers). Under Regulation 1(3), any loch, pond or reservoir is included, whether or not it discharges into a river or watercourse. The combined effect of these provisions is that "inland waters" in Scotland cover most fresh waters, but do not include any tidal waters seaward of the fresh-water limit. This means that, whereas the disposal of dredgings from estuaries and tidal waters in England and Wales benefits from exemptions, that from corresponding waters in Scotland does not.

3.5.3 The meaning of "agriculture"

The meaning of "agriculture" is taken from the Agriculture Act 1947, section 109(3), in England and Wales. In Scotland an identical provision in the Agriculture (Scotland) Act 1948, section 86(3), is used. "Agriculture" is defined in wide terms, and includes horticulture, fruit growing, seed growing, dairy farming and livestock breeding and keeping, the use of land as grazing land, meadow land, osier land, market gardens and nursery grounds, and the use of land for woodlands where that use is ancillary to the farming of the land for other agricultural purposes.

3.5.4 Conditions

The conditions under which the spreading of dredgings from inland waters on land used for agriculture is exempt from Waste Management Licensing are as follows:

• no more than 5000 tonnes of waste per hectare may be spread on the land in any period of 12 months. This figure refers to the weight of the waste as emplaced and not dry weight
• the activity must result in benefit to agriculture or ecological improvement
• where the waste is to be spread by an establishment or undertaking (i.e. an organisation rather than a private individual) that body, which will not necessarily be the owner or occupier of the land, must furnish the WRA for the area with the following particulars;

  - the establishment's or undertaking's name and address, and telephone or fax number (if any)
  - a description of the waste, including the process from which it arises
  - where the waste is being or will be stored pending spreading
  - an estimate of the quantity of the waste
  - the location and intended date of the spreading of the waste
  - any additional information required.

In a case where there is to be a single spreading, these particulars should be provided in advance of carrying it out. If, however, there is to be a regular or frequent spreading of waste of a similar composition, the particulars must be provided every six months. In that case, the estimate of the quantity of waste should relate instead to the total amount to be spread during that period, and the frequency of spreading should be substituted for the intended date. If a different type or quality of waste is to be spread, the particulars should be furnished in advance.

3.5.5 The meaning of "benefit to agriculture or ecological improvement"

The phrase "benefit to agriculture or ecological improvement" is taken from the EC Directive on Waste, Annex IIB R10, where it is used to describe one of the operations that may lead to the recovery of waste. No legal definition is provided either in the Directive or in the Waste Management Licensing Regulations. The phrase must therefore be interpreted in a commonsense way.

DoE Circular 11/94 states that "benefit to agriculture or ecological improvement" should be interpreted according to the type of wastes and category of land involved. It does not include bulk application simply to raise the level of land. It does include applications that convey positive benefits. Properly qualified advice will need to be produced stating the quantities and frequencies (application rates) that will result in positive benefits. The issues of benefit to agriculture and ecological improvement are further discussed in Sections 5.5 and 5.6.

3.5.6 Storage of dredgings

Where dredgings are intended to be spread on agricultural land in reliance on the above exemption, there also is an additional exemption in Schedule 3 paragraph 7(5) for their storage at the place where the spreading will take place. This does not apply to the storage of waste in liquid form unless it is stored in a secure container or lagoon and no more than 500 tonnes is stored in any one container or lagoon. A container or lagoon is considered to be secure if all reasonable precautions are taken to ensure that the waste cannot escape from it and members of the public are unable to gain access to the waste. The exemption does not, however, authorise the construction of a lagoon, and the word "lagoon" is not legally defined.

3.5.7 Screening and dewatering

Schedule 3 paragraph 25(5)(b) also exempts the treatment by screening or dewatering of dredgings prior to spreading them on agricultural land under the exemption in paragraph 7(1). For the exemption to apply, this treatment <u>must</u> be carried out either on the bank or towpath where the dredging or clearing takes place or at the place where the dredgings are to be spread. The meaning of "bank" is considered in Section 3.7.2.

3.6 LAND RECLAMATION

Schedule 3 paragraph 9 exempts the spreading of waste from dredging any inland waters on any land in connection with the reclamation or improvement of that land, provided that:

- by reason of industrial or other development the land is incapable of beneficial use without treatment
- the spreading is carried out in accordance with a planning permission for the reclamation or improvement of the land and results in benefit to agriculture or ecological improvement
- no more than 20,000 m³ per hectare of such waste is spread on the land.

Schedule 3 paragraph 25(5)(c) also exempts the treatment by screening or dewatering of such dredgings prior to their spreading under this exemption, and the treatment may be carried out either on the bank or towpath where the dredging operation takes place or at the place where the dredged material is to be spread. Both these exemptions are only available if the activity is carried out by or with the consent of the occupier of the land, or if the party doing it is otherwise entitled to do so (e.g. under statutory powers).

REGULATION

The purpose of this provision is to facilitate the reclamation of contaminated land. The Regulations do not state whether the requirement that land reclamation or improvement must be authorised by a planning permission is limited to situations where a planning application has been submitted and granted, or whether it may also include permitted development under a General Development Order (e.g. the spreading of dredged material by statutory undertakers: see Section 3.9.4). Since the Town and Country Planning Act 1990, section 63, treats permission granted by a development order as a planning permission, it would seem reasonable to adopt a similarly broad approach in this context. "Inland waters" and "benefit to agriculture or ecological improvement" have the same meaning as discussed in Section 3.5.

3.7   DEPOSIT ON BANKS AND TOWPATHS

3.7.1   Exemption

Schedule 3 paragraph 25 exempts the deposit of waste arising either from dredging inland waters or from clearing plant matter from inland waters in two situations. The first applies if the waste is deposited along the bank or towpath of the waters where the dredging or clearing takes place. However, if an establishment or undertaking deposits waste from dredging, this exemption is only available where the waste is its own. The second exemption applies if the waste is deposited along the bank or towpath of any inland waters so as to result in benefit to agriculture or ecological improvement. In either case, the total amount of waste deposited along the bank or towpath on any day must not exceed 50 tonnes for each metre of the bank or towpath along which it is deposited.

3.7.2   The meaning of "bank"

The geographical extent of a "bank" is not defined in the Regulations, and must be regarded as a question of fact to be decided according to commonsense principles in each situation. However, the Court of Appeal in the case of Jones v. Mersey River Board [1958] 1 QB 143 adopted a wide interpretation of the word "banks" in land drainage legislation, and held that it means "*so much of the land adjoining or near to a river as performs or contributes to the performance of the function of containing the river*". This means that a bank is not limited to a towpath, but can include some adjoining land associated with a river. It is likely to extend to land within the reach of an excavator, but would not include the whole of a large field. The WRA should be consulted in each case.

"Inland waters" and "benefit to agriculture or ecological improvement" have the same meaning as discussed in Section 3.5. The exemptions do not extend to waste deposited in a container or lagoon. However, they do permit the prior treatment by screening or dewatering of the dredgings or plant matter on the bank or towpath of the waters where either the dredging or clearing takes place or the waste is to be deposited. These exemptions only apply if the activity is carried on by or with the consent of the occupier of the land, or if the party doing it is otherwise entitled to do so.

3.7.3   Public rights of way

The deposit of dredgings on banks and towpaths alongside inland waterways may have significant implications for public rights of way (i.e. public footpaths, bridleways, etc.). Many rights of way follow banks or towpaths where the deposit of waste would be exempt from the need for a Waste Management Licence if the waste originates from the adjacent watercourse, or if the deposition results in benefit to agriculture or ecological improvement. In either case, the total amount of waste deposited must not exceed 50 tonnes per metre of bank or towpath per day.

The County Surveyors Society believes that a rate of deposition considerably below 50 tonnes per metre is likely to result in a substantial obstruction to members of the public wishing to use the right of way, unless consideration is given to the existence of the right of way, either by controlling the area over which deposit takes place or, if necessary, by means of a temporary diversion of the right of way. In either case the operator should ensure that any damage to the surface of a public right of way is properly reinstated.

Since the surfaces of all highways maintainable at the public expense are vested in the Highway Authority under Section 263 of the Highways Act 1980, it would appear that the establishment seeking to deposit dredgings may not have a legal occupation of the surface of the right of way consistent with the right to deposit waste. In most cases, however, liaison between the establishment involved and the Highway Authority could be expected to allow a satisfactory solution to be found. Operators should consult with the relevant Highway Authority in all cases where a public right of way is likely to be affected by the deposit of dredged material and should establish the location of rights of way prior to undertaking bankside disposal.

3.8 OTHER EXEMPTIONS

3.8.1 Introduction

There are six other categories of exemption in Schedule 3 of the Regulations which do not expressly refer to dredgings but are potentially relevant to them because of their broad subject matter. These are contained in paragraphs 7(2), 12, 19, 21, 40 and 41. However, it is arguable that the exemptions which specifically deal with dredgings from inland waterways should be applied where there are apparent overlaps between them and other exemptions. WRAs may therefore be reluctant to accept them.

3.8.2 Spreading waste soil, compost or plant matter on operational land

Schedule 3 paragraph 7(2) permits waste soil, compost or other plant matter to be spread on the operational land of a limited number of organisations (including internal drainage boards and the NRA, but not British Waterways) provided that no more than 250 tonnes are spread on the land in any period of 12 months and the activity results in benefit to agriculture or ecological improvement. The waste may also be stored at the place where it is to be spread. "Operational land" is a term defined in planning legislation, and refers to land used by statutory undertakers for the purpose of carrying on their undertaking.

3.8.3 Composting biodegradable waste

Schedule 3 paragraph 12 permits the composting of biodegradable waste at the place where it is to be used, or at any other place occupied by the person producing the waste or using the compost, provided that the total quantity of waste being composted at that place does not exceed 1000 m³ at any time. Biodegradable waste for composting may also be stored at the place where it is produced or is to be composted. "Composting" is defined as including any biological transformation process that results in materials which may be spread on land for the benefit of agriculture or ecological improvement. No statutory definition is provided of "biodegradable", which must therefore be interpreted on scientific rather than legal principles (see also Section 4.6).

REGULATION

3.8.4 Storage and use of waste from excavations for construction work

Schedule 3 paragraph 19 permits the storage and use on a site of waste arising from "excavations", provided that it is suitable for the purposes of "relevant work" which will be carried on there and, if it is not produced on the site, is not stored there for longer than 3 months before the work begins. "Relevant work" means construction work apart from land reclamation. It includes the deposit of waste on land, but only in connection with the provision of recreational facilities there or the construction, maintenance or improvement of a building, highway, railway, airport, dock or other transport facility on the land. The relevance of this exemption to the disposal of dredgings depends on whether dredging operations can properly be described as "excavations", and it is arguable that this term more naturally relates to land rather than waterways, although the point has not been tested in the courts. It is therefore essential to consult the WRA before relying on this provision.

3.8.5 Treatment of waste plant matter

Schedule 3 paragraph 21 permits the chipping, shredding, cutting or pulverising of waste plant matter on any premises, provided that those activities are carried out for the purposes of recovery or re-use and no more than 1000 tonnes of such waste are dealt with in any period of 7 days. The storage of up to 1000 tonnes of waste plant matter at any time in connection with this treatment (i.e. before, during or after it) is similarly permitted on the premises. "Premises" is a wide term, and may include buildings or open land. Activities "for the purposes of recovery or re-use" should cover preparatory acts such as pre-treatment as well as subsequent recovery, and there is no requirement that exempt pre-treatment must take place at the same site as recovery or be undertaken by the same person.

3.8.6 Temporary storage away from the place of production

Schedule 3 paragraph 40 exempts the temporary storage of non-liquid waste <u>away from the site where it is produced</u>. "Non-liquid" is not legally defined. The following conditions must be satisfied:

- the waste must be stored in a secure container or containers
- no more than 50 m^3 of waste may be stored at a time
- the period of storage must not exceed 3 months
- the owner of the container and the occupier of the land must consent to the storage
- the main use of the site must not be the reception of waste for disposal or recovery elsewhere (i.e. it should not be a transfer station)
- the waste must be stored for the purpose of collection or transport (i.e. not as part of a disposal or recovery operation).

3.8.7 Temporary storage at the place of production

Schedule 3 paragraph 41 exempts the temporary storage of waste <u>on the site where it is produced</u> pending its collection. This exemption should be wide enough to permit dredgings to be stored temporarily on banks or adjoining land, provided that the consent of the occupier is obtained. No time limit is specified, but the period of storage should not be excessive or indefinite. Additional conditions apply if the dredgings contain "special waste" as described in Section 3.3.

3.9 WASTE MANAGEMENT LICENCES

3.9.1 Introduction

Under section 33(1) of the Environmental Protection Act 1990, it is a criminal offence to deposit, treat, keep or dispose of controlled waste (now restricted to controlled waste that is also Directive waste as outlined in Section 3.2.1) on any land unless there is a Waste Management Licence in force authorising the activity on that site. It is similarly an offence to knowingly cause or knowingly permit such unauthorised activity by another party. This means that a contractor or employee as well as the party employing them are all potentially liable. While "deposit" must be interpreted in a commonsense way, and may be either temporary or permanent, the phrase "treat, keep or dispose of" is now limited to the waste disposal and recovery operations listed in the EC Directive on Waste and reproduced in Schedule 4 Parts III and IV of the Waste Management Licensing Regulations. Further changes in law can be expected in the future as a result of a new EC Directive on Landfill. This is likely to restrict the co-disposal of different types of waste, require the upgrading of existing landfill sites and introduce new criteria for the acceptance of waste.

The waste disposal operations potentially relevant to dredging include:

- tipping of waste above ground or underground (D1)[1]
- the land treatment of waste (e.g. biodegradation of liquid or sludge discards in soils, etc.) (D2)
- surface impoundment of waste (e.g. placement of liquid or sludge discards into pits, ponds or lagoons, etc.) (D4)
- storage of waste pending another listed waste disposal operation (apart from temporary storage, pending collection, on the site where the waste is produced) (D15).

Relevant waste recovery operations include:

- spreading of waste on land resulting in benefit to agriculture or ecological improvement (including composting and other biological transformation processes) (R10)
- storage of waste consisting of materials intended for submission to any listed waste recovery operation (apart from temporary storage, pending collection, on the site where the waste is produced) (R13).

Under the Controlled Waste Regulations 1992, SI 1992 No. 588, different types of "controlled waste" are divided into three categories as "household", "industrial" or "commercial" waste. According to Schedule 3 of those Regulations, controlled waste from dredging operations should be treated as "industrial waste". Industrial waste may, however, be either inert or biodegradable (see Sections 4.5 and 4.6).

Therefore, in situations where dredgings qualify as waste and there is no applicable exemption, a licence must be obtained not merely for their ultimate deposit on land, but also for any activity involving their disposal or recovery within the terms of the EC Directive. The maximum penalty for non-compliance is a fine of £20,000 and/or 6 months imprisonment on summary conviction in a magistrates' court, or an unlimited fine and/or 2 years' imprisonment on indictment in a Crown Court. Even when a licence has been granted, there remains the possibility of committing an overriding offence under section 33(1)(c) of the Environmental Protection Act 1990, if controlled waste is treated, kept or disposed of in a manner likely to cause pollution of the environment or harm to human health.

REGULATION

---

[1] The references in parentheses relate to paragraphs in Annex II of the EC Directive on Waste.

3.9.2 Application procedure

It is the responsibility of the person in occupation of the land on which the activity requiring a Waste Management Licence takes place to obtain it.   The application must be made to the WRA for the area in which the land is situated and, if the site crosses the boundary between two authorities, a licence must be sought from each. However, it is possible that they may agree to one WRA taking responsibility for a cross-boundary site. There is no prescribed common form of application under the Regulations, and WRAs are therefore free to issue their own documentation. Under section 36 of the Environmental Protection Act 1990, a licence may not be issued for a use of land requiring planning permission, unless that permission has been granted or a certificate of lawful use or development has been obtained from the planning authority. Otherwise, the only grounds on which a licence may be refused are pollution of the environment, harm to human health or (unless the activity is authorised by a planning permission) serious detriment to the amenities of the locality.

The requirements in this respect are:

* the WRA must notify the applicant of its decision within 4 months of receiving an application, or it will be deemed to have been refused

* the WRA and the applicant may agree in writing to extend the time limit for deciding an application

* the applicant may appeal to the Secretary of State within 6 months against a refusal to issue a licence or against the conditions included in one.

3.9.3 Statutory consultation

If a WRA proposes to issue a Waste Management Licence, it must first consult the NRA (or the River Purification Board in Scotland) and the Heath and Safety Executive, in order that the water pollution and safety implications can be assessed. In Scotland, a WRA that is not itself the district planning authority must also consult the general planning authority for the area. If the NRA or the River Purification Board disagrees with the WRA, the matter must be referred to the Secretary of State for the Environment, Wales or Scotland, whose decision is final.

If any part of the land is inside a Site of Special Scientific Interest (SSSI), the WRA must also consult English Nature, the Countryside Council for Wales or Scottish Natural Heritage, although there is no procedure for referring disagreements to the Secretary of State. However, if an application is likely to have a significant effect on a special protection area (SPA) or special area of conservation (SAC) designated under the EC Birds Directive 79/409/EEC or Habitats Directive 92/43/EEC, the Conservation (Natural Habitats, etc.) Regulations 1994, SI 1994 No. 2716 impose strict limitations on the ability of the WRA to grant a licence, and this applies even if the waste management site is located outside the protected area.

The potential statutory consultees in each country are therefore as follows:

England

* NRA
* Health and Safety Executive
* English Nature.

Wales

* NRA
* Health and Safety Executive
* Countryside Council for Wales.

Scotland

- River Purification Board
- Health and Safety Executive
- Local planning authority
- Scottish Natural Heritage.

A WRA may choose to consult other interested parties, such as the owner of any overhead power cables, as appropriate.

3.9.4 Planning permission

Planning permission is required for the disposal of dredgings if it involves the development of land. Under the Town and Country Planning Act 1990 and the Town and Country Planning (Scotland) Act 1972, development takes place if the use of land is changed to the deposit of waste, or if an existing waste disposal site is extended in area or raised above the level of the adjoining land. An application for planning permission must normally be submitted to the local planning authority, but some categories of development are automatically permitted by a General Development Order.

The Town and Country Planning (General Permitted Development) Order 1995, SI 1995 No. 418, Schedule 2, Part 17, Class D, grants automatic planning permission for statutory undertakers in England and Wales to spread dredged material on any land. Statutory undertakers are bodies including British Waterways, the NRA and harbour authorities, which are authorised by legislation to operate canals, inland navigations and harbours, etc. A similar planning permission is granted in Scotland by the Town and Country Planning (General Permitted Development) (Scotland) Order 1992, SI 1992 No. 223, Schedule 1, Part 13, Class 37, except that it is limited to the statutory undertakers' "operational land" (i.e. land which they hold for the purposes of their business) and so does not apply if they deposit dredgings on other land. However, under the Conservation (Natural Habitats, etc.) Regulations 1994, a planning permission granted by General Development Order which is likely to have a significant effect on an SPA or SAC (as described in Section 3.9.3) requires the written approval of the local planning authority. In addition, the Town and Country Planning (General Permitted Development) Order 1995 withdraws permitted development rights in England and Wales if a project needs Environmental Assessment (see Section 3.9.5). In that case, it will be necessary to submit a planning application. There is a statutory procedure under the Town and Country Planning (Environmental Assessment and Permitted Development) Regulations 1995 (SI 1995 No. 417), for obtaining an opinion from the local planning authority about the status of a proposed development in this respect.

3.9.5 Environmental Assessment

EC Directive 85/337/EEC requires certain projects that could have a significant effect on the environment to be subject to an Environmental Assessment (EA) procedure before permission for the development is given.   This Directive is implemented in the UK through the planning system by the Town and Country Planning (Assessment of Environmental Effects) Regulations 1988, SI 1988 No. 1199 and the Environmental Assessment (Scotland) Regulations 1988, SI 1988 No. 1221. Projects relevant to dredgings are listed in Schedule 2 of these Regulations. They are installations for the disposal of controlled waste, sites for depositing sludge and coast protection works. Applicants for planning permission for these developments must provide an Environmental Statement if the local planning authority considers that they could have significant environmental effects. Sites for the disposal of special waste (see Section 3.3) are listed in Schedule 1 of the Regulations, and must always have an Environmental Assessment.

REGULATION

Several public bodies including local authorities, the NRA (or River Purification Board), English Nature and the Countryside Commission (or Countryside Council for Wales or Scottish Natural Heritage) must be consulted before and after the preparation of the Environmental Statement. The Statement and the views of the statutory consultees must then be taken into account when the planning application is decided. In England and Wales, improvements to existing land drainage works (such as flood defences) are granted automatic planning permission under the General Development Order. However, Environmental Assessment of such developments is administered by the Ministry of Agriculture, Fisheries and Food (MAFF) instead of the local planning authority under the Land Drainage Improvement Works (Assessment of Environmental Effects) Regulations 1988, SI 1988 No. 1217. In Scotland, the Scottish Office has responsibility for the Environmental Assessment of drainage works. As explained in Section 3.9.4, automatic planning permissions granted to statutory undertakers by a General Development Order are withdrawn in other cases where Environmental Assessment is required by the EC Directive.

3.10   FIT AND PROPER PERSONS

3.10.1 Introduction

Under section 36(3) of the Environmental Protection Act 1990, a WRA must be satisfied that an applicant for a Waste Management Licence is a "fit and proper person" to hold it. Section 74 prescribes three tests for this purpose: relevant offences, technical competence and financial resources. In addition, Waste Management Paper No. 4 contains official guidance on the interpretation of "fit and proper persons", to which WRAs are statutorily required to have regard. The legal criteria are as follows.

3.10.2 Relevant offences

The WRA must take into account the fact that an applicant has been convicted of a "relevant offence", and this also applies to employees of the applicant and companies of which they are a director or officer. Regulation 3 of the Waste Management Licensing Regulations 1994 contains a list of criminal offences under environmental legislation which are to be treated as "relevant" in this context. The WRA has a discretion to reject the applicant because of a conviction.

3.10.3 Technical competence

Management of the prospective licensed activities must be in the hands of a "technically competent person". Regulation 4 of the Waste Management Licensing Regulations 1994 prescribes the technical qualifications which must be held in order to satisfy this requirement for different types of waste facility. These are certificates awarded by the Waste Management Industry Training and Advisory Board (WAMITAB). Regulation 5 contains temporary transitional provisions to safeguard the position of existing and experienced managers who lack the necessary qualifications. These concessions have been extended by the Waste Management Licensing (Amendment etc.) Regulations, SI 1995 No. 288, and the Waste Management Licensing (Amendment No. 2) Regulations 1995, SI 1995 No. 1950.

For most types of landfill operations (i.e. sites receiving special waste or biodegradable waste, or those which require substantial engineering works to protect the environment) a level 4 certificate is demanded. An exception is for landfill sites with a total capacity exceeding 50,000 m^3 which receive only inert waste. For these, a level 3 certificate is sufficient. In the case of sites of a type not covered by the WAMITAB scheme (e.g. small landfills taking only inert waste) Waste Management Paper No. 4 contains guidance on appropriate standards. No legal definition is provided in the Regulations to distinguish inert from biodegradable waste, and a scientific judgement must therefore be made by WRAs in order to identify the status of dredgings. However, the following definition of "inert waste", which appears in the Waste Management Licensing (Fees and Charges) Scheme 1995, may be of assistance:

- *"waste which, when disposed of in or on land, does not undergo any significant physical, chemical or biological transformation."*

The legislation is also unclear about the extent to which a licensed site must be under constant supervision by a qualified manager, although Waste Management Paper No. 4 accepts that more than one site may be under the day-to-day control of the same individual or group. The law is flexible about methods of achieving technically competent management, but it is the responsibility of applicants to satisfy the WRA in each case that their arrangements are adequate.

3.10.4 Financial resources

The applicant must make adequate financial provision to discharge the obligations arising from the licence. This requirement is not satisfied if the applicant either lacks the necessary resources or has no intention of making them available. Guidance on this issue is provided in Waste Management Paper No. 4.

3.11 DUTY OF CARE

Section 34 of the Environmental Protection Act 1990 imposes a "Duty of Care" on all holders of waste. This applies not only to waste managers who keep, treat or dispose of waste (for which a Waste Management Licence will be required), but also to those who produce, carry or import waste or have control over it as a broker. A producer of waste is a person who undertakes the works which give rise to it. Thus, a contractor who carries out dredging operations on behalf of another party will qualify as a "producer" of waste, and will be subject to the Duty of Care, even though he/she will not have to apply for a Waste Management Licence unless he/she is also the occupier of the land. The employer who engages the services of a contractor, on the other hand, may escape the duty of care if he/she does not "hold" the waste as described above, despite issuing the instructions that result in its creation. However, each case will depend on its individual circumstances, and it will not be possible to delegate responsibility in all situations. Discussion with the WRA is therefore highly desirable.

The "duty" is to take all reasonable steps to prevent the unauthorised handling of waste, to prevent its escape, and to ensure that it is only transferred to an authorised person, who must be provided with sufficient written information to enable him/her to comply with his/her own Duty of Care. Copies of the documentation must be retained by both parties for 2 years after the transfer of waste. The Duty of Care is transferable in the sense that it applies until the waste itself is transferred <u>in compliance with the prescribed legal formalities</u> to an authorised person, who then assumes the duty instead. Authorised persons include the holder of a Waste Management Licence and also registered carriers of waste. Professional carriers and brokers of waste are required to register with the WRA under the Control of Pollution Amendment Act 1989, the Controlled Waste (Registration of Carriers and Seizure of Vehicles) Regulations 1991, SI 1991 No. 1624, and the Waste Management Licensing Regulations 1994. The Secretary of State has issued a statutory code of conduct providing practical guidance on how to discharge the Duty of Care, which is legally admissible in court as evidence of compliance. However, the scope of the Duty of Care has not yet been fully tested in the courts, and the producers and holders of dredgings should be cautious about assuming that legal responsibility has been transferred. An updated code was published for consultation by the DoE in August 1995.

Because both the Duty of Care and the obligation to obtain a Waste Management Licence are statutory requirements, these responsibilities cannot be waived by private arrangements between employer and contractor. However, a contract may result in a dredging contractor becoming a producer of waste and thus subject to the Duty of Care. It is appropriate to ensure, as far as possible, that contractors observe their statutory duties by including a warranty in dredging contracts that they will conform with the provisions of legislation and indemnify their employer against any penalties or liabilities.

3.12 PROTECTION OF GROUNDWATER

Regulation 15 of the Waste Management Licensing Regulations gives effect to provisions of EC Directive 80/68/EEC on the Protection of Groundwater against Pollution caused by Certain Dangerous Substances. The dangerous substances to which the Directive applies are divided into two lists, which are set out in Box 3.12.1. "Groundwater" is defined in the Directive as *"all water which is below the surface of the ground in the saturation zone and in direct contact with the ground or subsoil"*.

According to Regulation 15, where a WRA proposes to issue a Waste Management Licence which might lead to a direct or indirect discharge of a List I or List II substance into groundwater, it must ensure that the intended activity is subjected to prior investigation, and must not issue the licence until it has checked that the groundwater will undergo the requisite surveillance. The information needed for these investigations will be demanded from the applicant, but the expert assessment of it will be undertaken by the WRA in consultation with the NRA (or RPB in Scotland). In practice, the NRA or River Purification Board will be likely to object at the planning stage if the applicant proposes to discharge List I substances to surface or groundwater from a waste disposal site, and will be unlikely to grant a consent to discharge under such circumstances. Subject to this, a WRA has the following powers under Regulation 15:

- in the case of a potential direct or indirect discharge of a List I substance, the WRA may issue a licence if the groundwater is already permanently unusable, provided that the substance will not impede mineral exploitation and all technical precautions are taken to ensure that it cannot reach other aquatic systems or harm other ecosystems. If the groundwater is not permanently unusable, a licence may only be issued subject to conditions that will ensure the observance of all technical precautions necessary to prevent discharges of List I substances into groundwater

- disposal activities that may lead to a direct or indirect discharge of List II substances may be licensed provided that conditions are included to ensure that all technical precautions necessary for preventing groundwater pollution by those substances are observed.

3.13 FEES AND CHARGES

Section 41 of the Environmental Protection Act 1990 empowers the Secretary of State to make a statutory scheme prescribing application fees and subsistence charges for Waste Management Licences, together with fees for applications to surrender or transfer licences or modify their conditions. These fees are payable to the WRA, but a scheme may also prescribe consultation fees to be paid by WRAs to the NRA (or RPBs in Scotland).

The current Waste Management Licensing (Fees and Charges) Scheme 1995 came into force on 1 September 1995. However, the reader should be aware that at the time of issuing this document a revised Waste Management Licensing (Fees and Charges) Scheme 1996 was being released. The annexes to this scheme list the sums payable in respect of the treatment, keeping and disposal of different types of waste. These are subdivided into bands according to the maximum annual amount of waste authorised by a licence, and thus relate to the potential rather than the actual amount of waste involved. However, there is scope for confusion between some of the categories of waste. Thus, Table 3 of the Fees and Charges, which relates to most forms of waste disposal, has higher charging bands for "industrial waste" than for "inert waste", and since waste from dredging operations is legally classified as "industrial waste" under the Controlled Waste Regulations 1992, WRAs may consider that the disposal of dredgings must always be charged on the industrial waste tariff even if they are actually inert. However, Table 3 Part A column (1) paragraph (d) of the Charging Scheme states that the fees and charges for industrial waste apply only to such waste as is not listed in other specified paragraphs. Since inert waste is included in paragraph (c), it can therefore be argued that the lower tariff should be charged for the disposal of inert dredged material.

List I

List I contains the individual substances which belong to the families and groups of substances enumerated below, with the exception of those which are considered inappropriate to List I on the basis of a low risk of toxicity, persistence and bioaccumulation.

Such substances which, with regard to toxicity, persistence and bioaccumulation, are appropriate to List II are to be classed in List II.

1 Organohalogen compounds and substances which may form such compounds in the aquatic environment
2 Organophosphorus compounds
3 Organotin compounds
4 Substances which possess carcinogenic, mutagenic or teratogenic properties in or via the aquatic environment[1]
5 Mercury and its compounds
6 Cadmium and its compounds
7 Mineral oils and hydrocarbons
8 Cyanides

List II

List II contains the individual substances and the categories of substances belonging to the families and groups of substances listed below which could have a harmful effect on groundwater.

1 The following metalloids and metals and their compounds:

1	Zinc	8	Antimony	15	Uranium
2	Copper	9	Molybdenum	16	Vanadium
3	Nickel	10	Titanium	17	Cobalt
4	Chromium	11	Tin	18	Thallium
5	Lead	12	Barium	19	Tellurium
6	Selenium	13	Beryllium	20	Silver
7	Arsenic	14	Boron		

2 Biocides and their derivatives not appearing in List I
3 Substances which have a deleterious effect on the taste and/or odour of groundwater, and compounds liable to cause the formation of such substances in such water and to render it unfit for human consumption
4 Toxic or persistent organic compounds of silicon, and substances which may cause the formation of such compounds in water, excluding those which are biologically harmless or are rapidly converted in water into harmless substances
5 Inorganic compounds of phosphorus and elemental phosphorus
6 Fluorides
7 Ammonia and nitrites

[1] Where certain substances in List I are carcinogenic, mutagenic or teratogenic, they are included in category 4 of this List.

3.14 PUBLIC REGISTERS

WRAs are required by section 64 of the Environmental Protection Act 1990 to maintain public registers of licences, applications, appeals, convictions and other information specified in Regulation 10, but excluding matters listed in Regulation 11, information affecting national security under section 65 of the Act and commercially confidential information under section 66. There is also a requirement under Regulation 18 to keep public registers of exemptions.

3.15 HEALTH AND SAFETY REQUIREMENTS

The obligation on WRAs to consult the HSE if they propose to grant a licence application (see Section 3.9.3) enables that body to submit objections on health and safety grounds, although it cannot prevent the issue of a licence and there is no dispute settlement procedure. However, Regulation 13 prohibits the inclusion of any conditions in a Waste Management Licence for the sole purpose of securing the health and safety of persons at work. The HSE must therefore rely on its regulatory powers under the Health and Safety at Work etc. Act 1974, which are applicable to workplaces in general. There are no special safety regulations relating solely to waste disposal or dredging, but the Control of Substances Hazardous to Health Regulations 1994, SI 1994 No. 3246, apply to sites handling hazardous wastes. Health and safety issues are dealt with further in Chapter 7.

3.16 TRANSITION FROM CONTROL OF POLLUTION ACT 1974

Existing waste disposal licences already granted under the previous Control of Pollution Act 1974, Part I, were automatically converted into Waste Management Licences by Section 77 of the Environmental Protection Act 1990, unless any activity under a Control of Pollution Act (COPA) licence became an exempt activity under the new Waste Management Licensing Regulations, in which case the licence ceased to exist. The same conditions which existed for a COPA licence will continue to exist with a Waste Management Licence. However, in certain specific cases (e.g. sites accepting waste oil and PCBs) with potential groundwater implications the WRAs may wish to review COPA conditions to bring them up to standards set by new legislation (e.g. Groundwater Directive). The holder of such a licence is treated as a "fit and proper person" unless his or her circumstances change (see Section 3.10). Applications for licences submitted before 1 May 1994 and appeals against decisions on them were determined on the basis of the old system.

Operations that did not need a licence before, but do so now, were exempted until 1 May 1995 under Schedule 3 paragraph 43 of the Waste Management Licensing Regulations (as amended), provided that they were carried out at the same site by the same operator immediately before 1 May 1994. The exemption must be registered with the WRA. If an application for a Waste Management Licence was submitted before 31 July 1995, the exemption remains in force until the application and any appeal is decided (or until the right to appeal expires after 6 months).

3.17 RELATIONSHIP WITH DEPOSIT OF WASTE AT SEA

The deposit of waste at sea is regulated by the Ministry of Agriculture, Fisheries and Food (MAFF) in England and Wales, and the Scottish Office in Scotland, under Part II of the Food and Environment Protection Act 1985. This Act provides a licensing system for the deposit of substances and articles from vehicles and vessels, etc. in tidal waters below the level of mean spring high water. Since waste management under the Environmental Protection Act 1990 extends to land above the low water mark of ordinary spring tides, there is a potential overlap between the two systems on the foreshore. However, duplication of control is avoided by Regulation 16, which excludes the recovery or disposal of waste from Waste Management Licensing if it is authorised under the Food and Environment Protection Act 1985. A licence for the deposit of waste at sea therefore takes precedence. There remains, however, the possibility that the deposit of material on a beach could cross the line of mean spring high water and thus require approval under both regimes. In this case, the two licensing authorities (MAFF and the WRA) would be expected to consult each other.

The EC Directive on Waste applies to territorial waters as well as land, and consequently Schedule 4 of the Waste Management Licensing Regulations modifies the Food and Environment Protection Act 1985 to incorporate the concept of Directive waste and provide for the preparation of offshore waste management plans. Under international law, the UK is also committed to taking all possible steps to eliminate and prevent pollution by dumping of wastes at sea. Consequently, when considering applications for marine disposal, the Government now takes into account the availability of alternative options, including the beneficial use of dredged materials. This will inevitably increase the pressure to dispose of dredgings on land.

3.18   CONCLUSION

The Waste Management Licensing Regulations 1994 have produced a complex legal and administrative framework for regulating the disposal of dredgings. Many of the problems are due to the difficulty of reconciling European Community law and British environmental legislation. Although conventional definitions and standards will be established with the passage of time, it is likely that only litigation will provide definitive answers to some questions of interpretation. In the meantime, the best way of eliminating uncertainties is through prior consultation between operators and WRAs.

REGULATION

REGULATION

4 Characterisation and classification of dredged material

4.1 INTRODUCTION

This chapter introduces the concept of characterising dredged material; gives a detailed account of, and guidance on, sampling procedures and analytical requirements; guides the reader through basic classification and interpretation principles; and suggests preferred practices that could be used to produce reliable and cost-effective results.

It should, however, be clear that there is no legally established or single widely accepted scheme for classifying dredged material in the UK and no legal requirement to do so. There is presently consultation on a National Waste Classification Scheme which will be introduced on a voluntary basis initially. The waste category will be subcategorised into inactive, other or contaminated classes; but as yet the criteria for this are not clear. The classification schemes outlined in this chapter do not represent legally binding, formal comparisons. Their value is in providing guidance to the user on the significance of any contamination present in a dredged material. WRAs will judge each case on a site-specific basis.

4.1.1 The need for characterisation

Consultation with WRAs, the NRA and other regulators during the preparation of this document has demonstrated that they consider that an adequate knowledge of the nature and characteristics of the dredged material (i.e. the waste) will be the first requirement of the process of selecting and/or approving a method of disposal. Simply describing the waste as dredged material will be insufficient.

This information is one of a number of factors used to determine the acceptability of a disposal operation, whether an exempt use or licensed disposal operation is under investigation. Because of the receiving site's influence on pollution control, understanding will also be required of the nature of the waste in relation to the receiving environment. Some characterisation of sediments may also be required to obtain licences for storage and dewatering.

4.1.2 Why characterise?

The characterisation of dredged material includes defining its physico-chemical nature, defining the potential levels of contaminating substances, and interpreting such values in terms of potential beneficial or adverse impacts on the receiving environment. At first this would appear to be complicated. However, there are certain procedures which can be used to characterise dredged materials in a cost-effective manner while gaining the optimal amount of knowledge about the material.

Characterisation helps to guide the operator on handling requirements and disposal opportunities and helps determine waste management options. The operator can identify, and prepare for, the particular needs associated with the disposal of contaminated sediment. It is not just the nature of the waste, but also the environmental consequences of its placement that have to be considered.

The operator can use the characterisation to demonstrate the sediment's suitability for a preferred disposal option, for example:

- to confirm that an agricultural benefit or ecological improvement can be achieved
- to demonstrate that there are no significant pollution problems, as part of a risk assessment in a working plan for a proposed new disposal area
- to show that the dredged material meets the WRA requirements as a permitted waste type for a licensed disposal site already in operation (e.g. landfill site).

Most importantly, characterisation of material will aid discussion and negotiation between the regulators and operators for both licensing and granting exemptions.

There are three important points to bear in mind throughout the characterisation process:

- the more consultation that is carried out at the start of a project, the easier it will be to target the sediment evaluation to the disposal operation
- the more thought and effort that the operator puts into proposals for a disposal operation prior to consultation, the more receptive the WRA will be to appraising such proposals
- once the dredged material has been adequately characterised, any further analytical requirements for a regularly dredged waterway may be much reduced.

4.1.3 What does characterisation involve?

Dredged material needs to be characterised in terms of its physico-chemical properties and chemical constituents. The range of parameters which can be considered to characterise the dredged material is diverse. Sediment analyses provide valuable data for assessing the risk of disposing of contaminated material and are likely to become an important part of the management of dredged material. For example, dredged material's contaminants may dilute or increase the contamination depending on the existing contaminant concentration of the receiving soils.

The nature of the dredged material and the receiving environment is likely to mean that evaluations are made on a case-by-case basis. The potential variability of dredgings along waterways should also be borne in mind. Variability can occur where the bankside activity changes, for example at locations adjacent to and downstream of old factory sites, gas works, other contaminated land sites and current or historical industrial effluent discharges.

4.1.4 Consultation

WRA representatives stress the overriding importance of early dialogue between themselves and the exemption notifier or licence applicant. They understand the need to keep operators' costs to a minimum by restricting expenditure on material characterisation. However, a thorough understanding of the nature of the dredged material is likely to be necessary. Early consultation between the parties concerned will help focus the scope of any sampling and analysis programme required, rather than allowing an operator to submit a licence application or exemption notification with too little, too much or irrelevant data on potential contaminants.

During initial investigations, any historical data and background information available should be reviewed (see Section 4.3.4). This review, which could be carried out in collaboration with the WRA, should help to determine those characteristics of the dredged material which may require assessment, thereby helping to meet the requirement for cost-effective environmental and pollution control.

Evaluations of the physical nature of the dredgings, contamination (if any), reasons for considering the material as inert or not biodegradable, etc. can be discussed with the WRA in the light of the characterisation results and possible disposal options. A WRA is likely to be more responsive to a proposal than to an unsupported request for advice and therefore the operator should enter into consultation with suggested ways forward for disposal, backed up by such information.

When characterising the material, it should be noted that, in some circumstances (i.e. when an operation crosses from one WRA's area to another's), more than one WRA will be involved. If a difference in opinion arises between the WRAs, then they may have to be dealt with separately if the differences cannot be resolved. A NAWRO sub-group is working towards closer co-operation between regulatory authorities, and this may be facilitated by the formation of the Environment Agency. A WRA may discuss characterisation requirements with the NRA, and possibly MAFF, English Nature, etc., particularly in situations where they may be concerned about maintaining environmental quality. This is likely to be the case irrespective of whether an exemption or a licence is being applied for and there are standard forms of guidance which will be referred to when evaluating a proposed disposal option. Examples of such guidelines are given in Box 4.1.4.

Box 4.1.4   Key guidance and reference documents for evaluating waste disposal options with regard to environmental interests

Agency	Guidance Document	Relevant Interest
DoE	Waste Management Papers Nos. 4/26/26A/26B/27	Waste facilities/landfilling wastes/monitoring landfill sites/design, construction and operation of landfills/ controlling landfill gas
DoE/Welsh Office/Scottish Office Environment Department	Environmental Protection Act 1990: Part II Waste Management Licensing. The Framework Directive on Waste. Joint Circulars 6/94 and 11/95.	Policy and guidance on Waste Regulations
DoE	Waste Management Licensing Regulations 1994 and Amendments 1995	Waste Regulations
MAFF	Code of Good Agricultural Practice for the Protection of Soil, 1993	Environmental protection of agricultural soils
MAFF	Code of Good Agricultural Practice for the Protection of Water, 1991	Environmental protection of water resources
NRA	Policy and Practice for the Protection of Groundwater, 1992	Environmental protection of groundwater resources
DoE	Guidance on the Assessment and Redevelopment of Contaminated Land. ICRCL 59/83	Contamination standards
DoE	Code of Practice for Agricultural Use of Sewage Sludge, 1989	Disposal to agricultural land

4.2   CHARACTERISING CONTAMINATION IN DREDGED MATERIAL

The particulate matter which comprises a waterway's sediment has two natural sources: the erosion products of rock and soil, and the matter generated within the waterway - typically comprising algal material and mineral precipitates. The "natural" particulate material at any location represents a background level of "uncontaminated" sediment.

These factors may cause variability in physical characteristics. However, human activities are such that most sediment is likely to be contaminated to some degree in urban environments, industrial areas or areas of intensive land use. Inputs of contaminating substances include sewage and industrial effluents from outfalls (point-source inputs), run-off from roads and agricultural land, substances from waterways users, and even atmospheric deposition (diffuse-source inputs).

The sediment's constituents, its oxidation state, the contaminants, and the system's pH exert a large influence on the levels of contamination which can concentrate in the sediment. Such factors are reflected in chemical standards for sediment contamination in classification schemes.

The identification of sediment constituents may be pertinent to demonstrating no contamination, identifying beneficial sediment characteristics (e.g. grain size, organic matter, nutrients for agriculture), and/or identifying whether the sediment is inert or biodegradable.

4.3   SAMPLING

4.3.1   Introduction

Sampling is required in order to identify the characteristics of the dredged material prior to its dredging and disposal.

4.3.2   Setting sampling objectives

The objective for dredged material testing is to obtain an accurate and comprehensive data set, in a cost-effective manner, which characterises a sediment as fully as is practical and satisfies the regulatory organisations' requirements in terms of decision-making for an exemption or licence. Sampling objectives and a sampling strategy can be submitted to the WRA for agreement prior to sampling being carried out.

4.3.3   Sampling strategy

A sampling strategy is essential in order to evaluate, cost-effectively, the necessary characteristics and constituents of the dredged material. When designing a sampling strategy many factors need to be taken into account including:

- historical data
- sediment type and variability (e.g. clay, silt, sand, gravel, etc), if known
- the number, location and distribution of sampling sites
- sampling collection, preservation, storage and transport procedures
- quality control/quality assurance
- disposal options (to be considered against the information obtained from the samples).

Any sampling strategy has to be considered in terms of meeting a WRA's licence application requirements, as well as taking into account the magnitude of the disposal operation and the economic constraints on the investigation. The strategy also has to be planned to take account of the maximum storage times for sediments prior to testing, the turn-around time of the laboratory analysis, and the possibility of resampling or needing extra sampling for more detail (e.g. leachate tests).

Sampling is the foundation upon which the analysis rests. Sampling inaccuracies can significantly affect the results of the final assessment. Consultation has indicated that the WRA will be looking for evidence of quality assurance, throughout the sampling and analysis procedures, to support the validity of any decisions they make under the Waste Management Licensing Regulations.

Box 4.3.3 summarises the main factors that need to be taken account of when an operator is planning a sampling programme.

Box 4.3.3   Sampling information

- Collect and collate all background information about the sampling site and likely nature of the material (physical and contamination). Identify sampling equipment required.

- Plan sampling in advance. Design around disposal objective. Agree locations and number of samples with WRA and other consultees.

- Identify trained personnel, or contractors, to carry out sampling exercise. Samples must be representative of prevalent sediment conditions. Accurate sampling is crucial to the validity and value of the results produced in the laboratory.

- The client commissioning the testing and the sampler/laboratory analyst contracted to carry out the work should discuss in advance the nature of the work and other factors such as site access, channel navigation and use, etc. This can save considerable time, effort and cost.

- All sampling equipment should be pre-cleaned (decontaminated) prior to the work. Care should be taken to avoid cross-contamination between samples. Sufficient sample volume should be collected to satisfy all analytical requirements, and allowances should be made for any extra or re-sampling requirements.

4.3.4 Using background information and historical data

Background information can often provide fundamental guidance in respect to the sampling plan, particularly during pre-application discussions with the WRA. An overview of the dredging method, sediment handling (dewatering, storage, etc.), the size of the dredging area, and the nature and depth of the sediment, will aid in determining the required number and location of samples, and the equipment necessary for sampling.

By reviewing the available historical data, existing and potential areas or types of contamination (or its sources) may be identified (see Figure 2.4.1). Various sources of information can be reviewed, including the dredging history of a waterway, and previously determined geotechnical, geochemical, hydrodynamic and sediment contamination data.

4.3.5 Selecting sampling points

Once the characteristics of the channel length to be dredged have been identified as far as possible, the location of the sampling sites and the number of samples required can be planned. Requirements in this respect should be agreed with the WRA. If little is known about a waterway's sediment, a preliminary "survey" can be carried out to give an initial indication of which contaminants (if any) are present in various locations of the dredging area. Such a survey may take the form of informed consultation and consideration of the adjacent bankside activities. This approach can prove more effective in identifying potential contamination than the analysis of composite samples from a range of waterway locations.

Sampling locations should not be focused on particular contaminated areas; the location and numbers of samples should be representative of the whole area to be dredged.

- Composite sampling

  Composite samples may be useful to characterise channel lengths to be dredged; however, the true maxima of contaminants will not be defined. Composite samples should only be used where they represent the mass of material that will be mixed during the disposal process. They could be used for open (undeveloped) stretches of waterway, but sampling requirements may vary where bankside activity (e.g. industry, agriculture, etc.) changes.

• Random sampling

The sampling pattern itself will be dependent on: the site and its environs, the shape and size of the dredged area, land-use activities (e.g. point sources of contaminants), sediment distribution, etc. Random sampling techniques allow for some error and the probability of missing areas of contaminated sediment. In waterways, the channel may be divided into cells and samples randomly taken within each cell. Random sampling does not take into account background information, historical data and the experience of the investigator, whereas the use of such information can help identify areas of contamination and define the boundaries of clean and contaminated waterway stretches (Warren, 1990). Random sampling may prove a useful approach for choosing sampling points if little is known about a waterway, or if the waterway's characteristics are believed to be uniform.

• Sampling for vertical variation

For deeper dredging, core samples may be required to identify vertical variability in sediment characteristics. Vertical definition may not be required if the sediment is known to be vertically homogeneous, or if previous data have demonstrated that any contamination is limited to surface sediments.

4.3.6 Selecting numbers of samples

There is no definitive UK guide to the number of samples required per area or volume to be dredged. In general, the number of samples required can be considered to be inversely proportional to the amount of known information, the confidence that can be placed on those data, and the suspected degree of contamination (USEPA and USACE, 1991; 1994). However, Table 4.3.6, based on Oslo Commission (1993) *Guidance on Dredged Material Assessment*, indicates the number of sampling stations that could be used to obtain representative data where the nature of the dredging area is reasonably uniform. These guidelines are typically used for large dredging schemes (e.g. port navigation channels) for disposal to sea.

Table 4.3.6   Oslo Commission (1993) guidance on numbers of sampling stations

Amount dredged (m³)	Number of stations
Up to 25,000	3
25,000 to 100,000	4-6
100,000 to 500,000	7-15
500,000 to 2,000,000	16-30
>2,000,000	extra 10/million m³

When selecting sample numbers the operator should evaluate the length of dredging relative to the surrounding environment, and should be guided by the following considerations:

• the more samples collected, the better the sediment will be defined
• the mean value of several measurements at a single location will give generally less variable results than individual sample values
• samples may require multiple analytical measurements as single values do not describe variability; replicates allow statistical representation
• the database must be sufficient to satisfy licensing regulators
• sample numbers have to fit within the temporal and financial targets of the project
• analytical laboratories tend to charge less per sample as the number of sample measurements to be made increases. If a channel is likely to be regularly dredged, and no significant change in contamination levels is foreseen, one large-scale sampling programme may be more cost-effective than several smaller programmes.

In general, for channels in rural locations, with a few adjacent land uses, one or two samples per km will probably be satisfactory. The main factors affecting the sample numbers required are likely to be the operator's disposal objective(s), changes in land use, inputs of contaminants, channel use, channel flow and hydrological conditions, physical channel characteristics, and sediment heterogeneity.

Where the length in question contains both rural and industrial land uses, separating the length may aid management of the sample numbers.

The number of samples required needs to be decided on a case-by-case basis. Except in extreme cases, the possible sample numbers for maintenance dredging could range from one composite sample per 5 km (minimum) to ten individual samples per km (maximum), dependent on the channel characteristics and adjacent land uses. More samples, and from greater depths, will be required for capital dredging schemes.

4.3.7 Sampling regime report

An initial report on a waterway's environs could be produced at this stage (e.g. a baseline review). Such a report should cover the relevant characteristics of the site and the materials to be dredged and disposed of, and should make suggestions on a sampling programme with particular regard to the contaminating substances which may or may not be present. Any existing sediment characterisation data should be included, as this could remove the need for testing the characteristics for which sufficient information is known.

The presentation of such a report to a WRA would facilitate discussion on the operator's preferred disposal option and the testing requirements that are felt necessary to accommodate this proposal.

4.3.8 Sample collection

The sampling apparatus must be capable of collecting representative samples of *in situ* sediment. Appropriate equipment is therefore required. Disturbance of a sample of previously undisturbed sediment may cause geochemical changes to the sediment which can significantly affect the analytical results. Unless the sediments are known to be vertically homogeneous, or the deepest sediments are known to be uncontaminated, samples should be taken to the planned depth of excavation. Great care needs to be taken to avoid any accidental sample contamination.

There are various techniques by which samples can be collected without significantly disturbing the sediment.

- Sampling by hand

  If the sample location is easily accessible, and only surface sediment is required, it may be possible to collect it by hand. Simple hand tools such as plastic scoops, trowels or hand augers can be used for this purpose. Current sampling practices include scraping a bucket across the channel bed and taking hand samples from hopper barges and backhoes. Although such methods are suited to sample collection in soft mud, silty clay and, in particular, stony sediment, they produce poorly preserved samples, unrepresentative of their *in situ* condition. Hand samples should only be taken from a backhoe if the sediment is consolidated and a sufficient quantity is undisturbed. Ring samplers can be used to obtain samples from exposed sediment (e.g. intertidal areas at low tide). It is important to rinse and clean all devices between samples to prevent any cross contamination.

  Hand sampling is the most suitable approach where sediment cannot otherwise be collected by grabs or corers (e.g. for some stony sediments or unconsolidated fines).

- Grab samplers

   Grab samplers may be used to collect surface samples, rather than hand sampling. A grab sampler is typically used to collect small undisturbed samples from the sediment surface (down to about 15-20 cm) in a triggered bucket system. In order to penetrate different sediments the grab may have to vary in weight (e.g. 35-40 kg mud and 70-100 kg for sands). Common types of grab samplers include Van Veen, Ekman, Shipeck, Day and Peterson grabs.

   The use of a grab is preferable where surface sediments have not been previously disturbed by boat-traffic or heavy flows.

- Corers

   Corers are essentially hollow tubes, often made of Perspex, with a steel cutting edge which give cylindrical samples of sediment. Corers can be operated from a boat using a winch, from a bridge, or can be used obliquely from the bank. A simple gravity corer is allowed to fall under its own weight through the water and into the sediment. On impact the cutting edge penetrates the sediment, which fills the tube. In clayey sediments the depth of penetration using a simple gravity corer is around 1 m. Care must be taken to ensure that sampling does not unduly compact the sediment when inserting the corer. The corer is retrieved and the sample removed. Certain corers are not suitable for the sampling of uncohesive sediments such as sands and gravels. Common types of corers include vibro-corers, gravity corers and Craib corers.

   Corers are usually preferred to grab samplers for recording any sediment stratification and should be used where samples from depths between 0.15 and 2.0 m are required. For sediment penetration greater than 2 m, a vibratory or piston corer should be used.

- Boreholes

   Capital dredging operations often require the collection of sediment to gain geotechnical information from depths greater than 1 m. Boreholes are sunk to obtain geotechnical samples and at the same time sediments can also be taken for physico-chemical analyses.

It is worth remembering that the results of a sediment analysis are only as good as the sample itself. How the sample is collected is crucial to the accuracy of the final results. The WRAs have indicated their desire to see some form of quality assurance for sampling (as well as analysis).

4.3.9 Handling, preservation and storage

Sediment samples are subject to physical, chemical and biological changes as soon as they are taken from their *in situ* position. Appropriate handling, preservation and storage methods should be used by the sample collector to ensure that the sample composition is not altered to give unrepresentative data (USEPA and USACE, 1991; 1994).

1. Handling

The sample volume to be collected must satisfy the requirements of the individual testing objectives. Sufficient volume is needed to provide the sample for various contaminant analyses and storage requirements, and to retain portions for any later analysis that may be called for. A typical volume of sediment required for the testing of a wide range of chemical parameters is around 2-2.5 litres of material. When commissioning a sampling and analysis scheme, the operator should ensure that the analytical contractor provides suitable containers for the analytical requirements.

2. Record keeping

All samples should be recorded in a log book containing relevant details of the sampling exercise. The information should be recorded so that the record can provide quality assurance details (e.g. to the WRA), and to enable easy identification of all samples at a later date (e.g. when in the laboratory). The following information is required:

- time and date of sample
- sample identification code
- sample location
- depth from which sample taken
- initials of personnel
- brief description.

A location map of the sampling points is important. This should be drawn up before the sampling is carried out and amended, if necessary, during the survey. This information will assist the sampling personnel to locate the sampling points, and act as a useful reference diagram. A sample point location map will also provide consistency between sampling exercises.

Brief descriptions of the samples should also be recorded in order to aid sample identification and characterisation of the material. Details useful for this purpose include:

- sample colour
- sample odour
- sample structure (e.g. clay, silt, sand, gravel, etc.)
- any important features (e.g. oil films, alien material, debris).

3. Preservation and storage

If the operator intends to collect his/her own samples, then preservation and storage techniques should be discussed with the analytical contractor before sampling takes place. The contractor will be able to provide information, advice and equipment, for example sample containers with added chemical preservatives. Alternatively, the contractor will undertake the sampling as part of the contract with the necessary preservation methods. Samples should be placed in clean containers, which can be provided by the laboratory by arrangement. They should be delivered as soon as possible after collection to the laboratory.

4. Transport

Listed below are some important principles to take into account when moving the dredged material samples from the waterway to the laboratory.

- all sample containers should be labelled and consistent with the log book records
- the time between sample collection and analysis should be kept to a minimum; the greater the time elapsed, the less reliable the results
- the samples should be handled and stored in such a manner to ensure no cross-contamination
- samples should be stored and packed so as to ensure no damage or spillage takes place during transport
- samples which are potentially hazardous should be couriered, not posted.

4.4   SEDIMENT ANALYSIS

4.4.1   Introduction

Reliability and cost-effectiveness are likely to be the main concerns of any analytical programme. The following sections outline some important considerations when establishing an analytical programme.

Where a proposed disposal site is known, an operator should direct (as far as possible) the analysis to deal with the known characteristics of the receiving site. For example, if the area lies within a groundwater Source Protection Zone (NRA, 1992), leachate testing is likely to be required by the NRA. Also, if a disposal operation involves an exempt activity, for example as a benefit to agriculture, sediment testing may be required to prove the benefit and ensure no disbenefits.

4.4.2   Physical sediment analysis

When assessing the quality of dredged material and the implications for disposal, factors other than contamination also need to be taken into account. Physical parameters (e.g. grain size distribution, organic matter content, etc.) can be important in evaluating any contamination and in gaining an overall picture of the geochemical characteristics of the material. The physical make-up of the sediment also has an important influence on the behaviour of contaminants after disposal (e.g. contaminant mobility).

Grain size (or particle size distribution) can have important agricultural properties, particularly in terms of soil structure and hydraulic conductivity. Grain size analysis records the frequency distribution of the sediments' particle size range. The grain size classes used to describe the physical make-up of sediment are generally as follows:

- clay   <0.002 mm
- silt   0.002 - 0.06 mm
- sand   0.006 - 2.0 mm
- gravel   2.0 - 60 mm
- cobbles 60 - 200 mm
- boulders >200 mm

Source: BSI (1981)

There is an inverse relationship between a sediment's grain size and the surface area which influences contamination. That is, the smaller the grain size, the greater the surface area on which contaminants can be concentrated. The "sediment matrix" (i.e. grain size distribution) effect needs to be recognised when assessing the implications of dredged material disposal. For this reason usually either the sand fraction (i.e. material <2 mm) or possibly the silt (<64 µm) or clay (<2 µm) fractions are used for chemical analysis. For example, sediment of <2 mm is used for leachate testing.

Total solids are determined after the sample has been dried at a specific temperature. The total solids values are generally used to convert concentrations of contaminants and other chemical parameters from a wet-weight to a dry-weight value.

4.4.3   "Total" and "available" chemical contaminant concentrations

In some circumstances regulatory authorities may request two analyses to be undertaken for certain metals, the distinction being either "total" or "available". Total values will represent that concentration of a substance present in a sample. However, only a fraction of the total metal concentration will be ecologically significant. This is known as the available concentration (i.e. that amount of metal available for accumulation in the biota).

One approach to testing is to use the results of total analyses as a basis for identifying those metals that may require additional analysis for available concentrations. For example, a laboratory can be instructed to carry out available tests if concentrations of total metals are higher than acceptable levels (e.g. trigger concentrations). Such possibilities should be discussed with the WRA as early as possible during sampling and analysis strategy planning.

4.4.4 Contaminant mobility: leaching tests on dredged material

Consultation identified the NRA's prime concern in respect of environmental protection as potential contaminant movement to surface water and groundwater. The delineation of Source and Resource Protection Zones is a major component of the NRA's *Policy and Practice for the Protection of Groundwater* (NRA, 1992) and is important when assessing the risk to groundwater resources from various activities, including waste disposal operations. The NRA is currently producing maps which will identify (in relation to overlying soil properties) the vulnerability of the underlying groundwater. It will provide these maps to the WRAs for reference when evaluating applications under the Waste Management Licensing Regulations.

Once a potential receiving site has been identified, the applicant should consult the NRA (normally through the WRA) on the suitability of the site and the potential for contaminants being leached from the dredged material. The so-called "leachate tests" can form the basis of an assessment of the potential for contaminants to move from the dredged material through the soil and into the groundwater.

Total contamination values for the dredged sediment may give a general indication of any pollution potential, but the NRA is likely to require more detailed analyses to specifically determine contaminant mobility at the proposed disposal site. There are several different analytical leachate tests. The method most widely recognised for estimating the mobility of metals leached under the influence of rain is the DIN 38414 test (DIN, 1984). This test has been cited in the proposed EC Directive on Landfill of Waste.

The NRA has recently produced its own leachate test and is recommending its adoption as the standard method used to assess the risk of leachable contaminants. The test procedure is described in the interim NRA publication *Leaching Tests for the Assessment of Contaminated Land* (NRA, 1994). However, at the time of writing, the test has yet to be validated. The NRA stress that this "*is the basic test, designed to provide as much information as possible on largely inorganic (e.g. metals) contaminant leachability within certain time and cost constraints*" (NRA, 1994). The NRA have also indicated that more complex methods may be necessary to provide detailed leachate information. This would be particularly relevant to organic contaminants (e.g. PAHs, PCBs, pesticides, etc).

It is important to consult with the NRA and WRA to agree appropriate leachate tests and to ensure that they are representative of actual leaching potential in the field. Other factors may be relevant, such as the water quality of the waterway and water quality at any existing or prior dredging sites.

4.4.5 Chemical sample preparation and analytical methods

Sample preparation includes ensuring that all the contaminants are extracted from the dredged material. In general, inorganic substances (e.g. metals) are digested in acid to release into solution the substances bound to the sediment. For organic compounds the chemicals have to be extracted from the sediment and dissolved using other organic solvents or acids.

Detailed guidance on analytical procedures is given in the individual methods set out under the Standing Committee of Analysts' *Methods for the Examination of Waters and Associated Materials* (MEWAM; Standing Committee of Analysts, 1981). These are the standard analytical methods recognised in the UK. Consultation, however, established that some MEWAM methods are inappropriate, or have been replaced by more accurate and specific tests for certain contaminants and sediment types. An accredited laboratory should be able to provide advice on relevant methods. The various analytical techniques involve sophisticated equipment for the measuring of specific chemical constituents. Clients should also ensure that they know what methods of preparation and analysis a laboratory is using, and that these are appropriate. The laboratory contracted should be accredited for all analytical procedures.

4.4.6 Quality control and laboratory protocol

The determination of low concentrations of substances in dredged material needs to be consistent and accurate. Concern over analytical reliability has been voiced by analytical chemists and within certain Government departments, and has become more widespread with the recent growth of a competitive environmental testing industry (Rix, 1994). Consultation with regulatory authorities has indicated that some form of quality assurance will be required to demonstrate the reliability of the testing results.

In the UK there are several quality assurance schemes which address laboratory analysis. Such schemes are typically based on laboratory accreditation. The main laboratory accreditation schemes for sediment/soil analysts in the UK are the National Measurement Accreditation Service (NAMAS) and the British Standards Institution's BS 5750. Before contracting a laboratory to carry out any sampling and analysis, details of its quality control schemes should be obtained and reviewed. It is important to check that the laboratory is specifically accredited for sediment analyses.

The National Physical Laboratory runs NAMAS: the assessment, monitoring and accrediting of analytical laboratories. NAMAS is a voluntary scheme which evaluates the technical quality of laboratories' methodology including quality assurance/control, instrumentation, calibration and staff training. It is the primary laboratory accreditation scheme in the UK and is gaining increasing recognition around the world. However, it is worth noting that NAMAS accreditation is no absolute guarantee of quality. There are several other external accreditation schemes which may be an appropriate indicator of quality control and should be considered when selecting a laboratory to undertake a contract.

BS 5750 does not directly approve the analytical methodology and practice covered under NAMAS, but provides the quality management system for organisations supplying products or services.

The use of a laboratory with a quality accreditation scheme would be seen by the WRAs as a means of assurance that the results gained from an agreed analytical programme were of satisfactory accuracy. For this reason, the dredging operator must check that an analytical contractor has the appropriate quality control and laboratory accreditation for sediment testing.

4.5 CLASSIFICATION OF DREDGED MATERIAL AS "INERT" WASTE

4.5.1 Introduction

Concern has been expressed by waterways operators regarding the possible categorisation of licensed waste from dredging operations as industrial waste (see Section 3.13). This definition was last formally presented in the Controlled Waste Regulations 1992. The Waste Management Licensing Regulations 1994 do not include any schedules of industrial, commercial or household waste, so their definition can be assumed to remain as per 1992. Although dredged material has been categorised as industrial waste, this does not necessarily mean it is automatically considered to be contaminated to a greater or lesser degree than commercial or household wastes.

4.5.2 Defining inert waste

Under the Waste Management Licensing (Fees and Charges) Scheme 1995, inert waste is defined as "*waste which, when disposed of in or on land, does not undergo any significant physical, chemical or biological transformation*". Therefore there is likely to be some dredged material which is inert, although it is also classified as industrial waste. Discussions with various WRAs have revealed, however, wide ranging interpretations of how dredged material, classified as industrial waste, can nevertheless be proved to be inert.

It is apparent that, because of its industrial classification, some WRAs may consider that dredged material cannot be inert or non-biodegradable based on the assumption that at least a proportion of the dredgings will undergo a significant transformation. Conversely, other WRAs recognise that, although dredgings have been classified as industrial waste, there may be some cases where the dredged material will meet the above definition of inert waste. Gaining WRA acceptance that a dredged material is inert is likely to require analytical information and is likely to be considered on a site specific basis (see Box 4.5.2).

Box 4.5.2   WRA approaches to defining "inert"

For the purpose of Fees and Charges associated with the 1994 Regulations, two county authorities have classified dredgings to be deposited at British Waterways' licensed sites as inert.

Authority 1 has determined that all dredgings from British Waterways are inert.

Authority 2 has used levels of contamination to define inert, but this has not excluded British Waterways' highest contamination class material from being defined as inert.

Parameter	Maximum concentration notified to the authority	
	Authority 1	Authority 2
pH (range)	5-7	6-7
Antimony	11	31
Arsenic	87	341
Barium	331	502
Beryllium	2	4
Boron	48	22
Boron (available)	6	4
Cadmium	3	214
Chromium	68	1445
Cobalt	18	39
Cyanide	2	1
Lead	464	1386
Mercury	42	11
Molybdenum	2	6
Nickel	103	430
Organic matter (%)	11	27
PAH	73	5148
Phenol	15	8
Selenium	2	4
Silver	<5	5
Tin	53	261
Thallium	<5	<5
Vanadium	40	55
Zinc	808	1983

All concentrations in mg/kg unless stated otherwise. All parameters are total concentrations unless stated otherwise. No single sample has all the maxima listed above.

In addition, Berkshire County Council has stated the following in its Deposit Draft-Waste Local Plan (December 1994):

"*Dredgings from inland waterways are classed as <u>inert wastes</u> and generally do not create a disposal problem in Berkshire except for dredgings from the River Thames.*"

In the UK there is no formally adopted definition of inert in terms of contamination. However, the proposed EC Directive on the Landfill of Waste (COM (93) 275), 1993, incorporates a set of assignment values which can be used to characterise waste as inert for the purpose of landfilling. The assignment values are set for eluates (i.e. solutions of substances washed from the sediment; a laboratory equivalent to a leachate). Such eluate criteria are based on leachate testing, following the DIN (1984) analytical method. The proposed Directive states that *"Wastes whose eluate concentration is not above the maximum values fixed for inert wastes will be considered as such"*.

The National Waste Classification Scheme proposes to classify material as inactive, rather than inert. Inactive materials include "1.4.2 silt and dredgings", with dredged sand, gravel, clay or rock being included in "1.1 Naturally occurring rocks and sub-soils".

4.5.3 Proving inert waste for disposal

To establish whether an operator can dispose of dredgings as inert waste, the WRA will have to be assured that the material satisfies the physical, chemical and biological criteria. The water content of dredgings has been seen as precluding them from being classified as inert. Water content can cause difficulty in handling the materials and its reduction after deposition can be viewed as a significant physical transformation. Despite the classification of uncontaminated dredgings as inactive in the proposed National Waste Classification Scheme, some degree of dewatering may be required prior to its disposal.

As there is no existing legislation defining chemical inertness in terms of contamination, an operator could use the Kelly values to show the dredged material to be uncontaminated for guidance purposes, although there are no leachate values. It is likely that any indication of human-sourced contamination would render the inert classification of sediment untenable. However, if sediment analyses characterise the dredgings as "uncontaminated" (i.e. less than the Kelly values), then the operator could approach the WRA with a claim for chemical inertness on such grounds.

The current DoE guidance set out in Waste Management Paper No. 4 establishes a cautious precedent for WRAs when considering whether a waste is inert. For example, sites licensed for inert waste should only receive waste which can be considered "genuinely inert", particularly with regard to slowly biodegrading material. The working plan for a licensed site should therefore *"indicate the system for ensuring that only genuinely inert wastes are accepted. This system will usually entail assessment of the waste source in combination with checking the waste at the site"* (DoE, 1994b). It is likely that the licence conditions for a disposal site will specify the individual types of waste that are acceptable.

4.5.4 Conclusions

In conclusion, it is not possible to state categorically that any dredged material will be acceptable to the WRA as inert waste as there are no quantifiable definitions for evaluation purposes. Different WRAs have differing interpretations of the nature of dredged material, particularly in light of the guidance set out in Waste Management Paper No. 4. It is therefore important to consult the relevant WRA over the definition and acceptability of a dredged material as inert. Testing is likely to be required to prove such a case and results should be presented to the WRA for discussion. Regular communication is vital to efficiently and effectively resolving this matter.

4.6   CLASSIFICATION OF DREDGED MATERIAL AS "BIODEGRADABLE" WASTE

4.6.1   Introduction

"Biodegradable" refers to that part of the dredged material subject to biological decay (i.e. the free organic fraction). It is the biodegradation of organic material which leads to the generation of gases within a disposal site. Some organic materials are more readily decomposed than others and it is these that are of primary concern. Such materials may include food, animal or vegetable waste, sewage, and other organic sludges. Other materials, such as organic solvents, are less readily biodegraded and are likely to be a less significant source of biogenically generated gases (Butterworth, 1991).

It is important for an operator to establish whether the dredged material is biodegradable as this will affect the type of landfill/disposal site to which the waste can be deposited. The type of waste and its associated environmental risk are reflected in the Fees and Charges Scheme (DoE, 1995c) for waste management licensing. Disposal facilities can also be grouped according to their design and operation in terms of containing certain contaminants. Sites for contaminated material are designed to allow no significant migration of leachates or generation of landfill gas. Other sites do not have such contaminant measures if the waste type and environmental setting do not pose a significant environmental threat. Again, this is reflected in the disposal costs including fees and charges, as well as site construction (e.g. using a liner) and on-going monitoring costs.

It is therefore necessary to evaluate the amount of biodegradable material within a waste in order to assess the potential risk from gas production. Gases, principally methane (CH_4) and carbon dioxide (CO_2), are predominantly produced by anaerobic biodegradation. CO_2 can also arise through aerobic processes, particularly biotic respiration. Most concern over gas generation has evolved from the use of landfill sites and the potential explosive nature of the gases produced, as well as the health risk this has for site operators and users.

4.6.2   Proving (non-) biodegradable waste for disposal

In theory there is likely to be some form and degree of biodegradable material in all dredged material. In practice, however, there will be some circumstances when this is not significant. Waste Management Paper No. 26A (DoE, 1994c) identifies a measurement of less than 10% volatile solids as the level at which methane production is not significant.

There is debate and a general lack of research about the minimum amount of organic material within a waste which could give rise to the gas trigger levels of CH_4 1% volume/volume and CO_2 1.5% volume/volume. Therefore a specified definition of what value of organic matter constitutes biodegradable may be considered unreliable. However, Waste Management Paper No. 27 (DoE, 1994d) states that "*A gas monitoring programme should be incorporated into the design of all operating and proposed landfills irrespective of waste type.*" This will be considered by the WRAs during licensing and thus gas monitoring may be required for a licensed disposal site regardless of whether the dredgings contain a low proportion of biodegradable matter.

As discussed earlier in this section, inert waste is defined under Waste Management Licensing (Fees and Charges) Scheme as "*waste which, when disposed of in or on land, does not undergo any significant physical, chemical or biological transformation*". Given the definition of inert waste and the fact that biodegradation involves "a biological transformation", it follows that if a waste material is inert, it cannot undergo a significant biological transformation. Under the proposed EC Landfill of Waste Directive the biodegradable component of a dredged material would be represented by the total organic carbon. However, the EC Directive has not been formally implemented. It is questionable whether an eluate measurement of total organic carbon is sufficiently accurate to represent the material's biodegradable fraction. It is likely, therefore, that a WRA will require more evidence than would be provided either by the requirements of the EC Directive or by a volatile solids content of less than 10% as a sufficient description of the nature of the organic matter in dredged material (i.e. before accepting that a waste is not significantly biodegradable). However, such values, particularly the volatile content set out in Waste Management Paper No. 26A, establish an initial definition from which consultation can be undertaken with the WRA.

Waste Management Paper No. 27 defines biodegradable material as "*any organic matter that can be decomposed by micro-organisms*". Biodegradable material that may be present in dredgings includes animal and vegetable matter, sewage effluents, wood, etc. Therefore, if the dredgings to be disposed of contain plant matter, for example from leaves from trees overhanging a channel, or from an overgrown channel, the material is likely to be considered biodegradable.

Other factors affect the rate of biodegradation to produce gases. The optimal production of methane requires a pH between 6.5 and 8.5. Most dredged material will have a pH value within this range. Perhaps a more important factor is the role of the moisture content of the dredged material. Higher rates of gas production occur in moist disposal environments. Incoming refuse has an average moisture content of 25%. Even after dewatering, some dredged sediments will have a high moisture content. Coarser sediments such as sands and gravels will dewater better than clays and silts, which can take long periods of time to drain, even under specific treatment techniques.

4.7 CLASSIFICATION OF CONTAMINATION IN DREDGED MATERIAL FOR DISPOSAL PURPOSES

4.7.1 Introduction

There are several classification systems in current use which can be used to guide an exemption/licence applicant on the status of their dredged material in terms of contamination. None of these classifications are established statutorily in the UK for the classification of contamination in dredged material for disposal purposes. It is essential to recognise that these systems are used for guidance rather than definition. The individual nature of the receiving environments will be too variable to give tight definitions on where a certain level of a contaminating substance will have a significant impact and WRAs will judge each case on a site-specific basis.

The general classification schemes discussed below are derived from various research for different purposes. Only the British Waterways (BW) system is directly relevant to the disposal of dredgings from inland waterways, and specifically BW waterways; however, the Kelly values provide guidance on the classification of contamination in wastes (e.g. dredgings).

4.7.2 The development of characterisation and classification

It is important to understand the background to and basis on which any characterisation or classification system is established. The definition and interpretation of sediment characteristics involves an understanding of the interactive geochemical processes and reactions (characterisation). However, for the dredging and disposal of potentially contaminated material, a more simplistic view of the system's behaviour needs to be adopted. It is for this reason that classification schemes have been developed. Classification tables represent a basic method for identifying the significance of any contaminants in the waste/dredged material. To a certain extent they can be used to simplify assessments, as well as to guide (but not necessarily define) an approach to the disposal of inland waterway sediments. It is important to realise that the classification schemes should be used for guidance purposes only.

4.7.3 Kelly/Greater London Council Waste Disposal Authority guidelines for contaminated soils (Kelly, 1979)

Kelly's (1979) paper on *Site Investigation and Materials Problems* includes guidelines for data interpretation for various soil contaminants and the hazards they may pose for the development of contaminated land. Kelly suggested a range of values which may be considered typical for uncontaminated soils, slight contamination, contaminated soils, heavy contamination and unusually heavy contamination. These values have been based on contaminant effects in terms of health risks and environmental hazards including direct ingestion, inhalation, indirect ingestion, phytotoxicity, dermatitis problems and effects on fauna. For example, Kelly noted that the risks of cadmium are associated with its uptake into vegetables whilst boron and nickel are of more concern due to their phytotoxic effects. The Kelly guideline values are presented in Table 4.7.3.

Table 4.7.3   Guidelines for contaminated soils - suggested range of values (mg/kg on air dried soils, except for pH)

Parameter	Typical values for uncontaminated soils	Slight contamination	Contaminated	Heavy contamination	Unusually heavy contamination
pH (acid)	6-7	5-6	4-5	2-4	> 2
pH (alk)	7-8	8-9	9-10	10-12	> 12
Antimony	0-30	30-50	50-100	100-500	> 500
Arsenic	0-30	30-50	50-100	100-500	> 500
Cadmium	0-1	1-3	3-10	10-50	> 50
Chromium	0-100	100-200	200-500	500-2500	> 2500
Copper (avail)	0-100	100-200	200-500	500-2500	> 2500
Lead	0-500	500-1000	1000-2000	2000-1.0%	> 1.0%
Lead (avail)	0-200	200-500	500-1000	1000-5000	> 5000
Mercury	0-1	1-3	3-10	10-50	> 50
Nickel (avail)	0-20	20-50	50-200	200-1000	> 1000
Zinc (avail)	0-250	250-500	500-1000	1000-5000	> 5000
Zinc (equiv)	0-250	250-500	500-2000	2000-1.0%	> 1.0%
Boron (avail)	0-2	2-5	5-50	50-250	> 250
Selenium	0-1	1-3	3-10	10-50	> 50
Barium	0-500	500-1000	1000-2000	2000-1.0%	> 1.0%
Beryllium	0-5	5-10	10-20	20-50	> 50
Manganese	0-500	500-1000	1000-2000	2000-1.0%	> 1.0%
Vanadium	0-100	100-200	200-500	500-2500	> 2500
Magnesium	0-500	500-1000	1000-2000	2000-1.0%	> 1.0%
Sulphate	0-2000	2000-5000	5000-1.0%	1.0-5.0%	> 5.0%
Sulphur (free)	0-100	100-500	500-1000	1000-5000	> 5000
Sulphide	0-10	10-20	20-100	100-500	> 500
Cyanide (free)	0-1	1-5	5-50	50-100	> 100
Cyanide (total)	0-5	5-25	25-250	250-500	> 500
Ferricyanide	0-100	100-500	500-1000	1000-5000	> 5000
Thiocyanate	0-10	10-50	50-100	100-500	> 2500
Coal tar	0-500	500-1000	1000-2000	2000-1.0%	> 1.0%
Phenol	0-1	2-5	5-50	50-250	> 250
Toluene extract	0-5000	5000-1.0%	1.0-5.0%	5.0-25.0%	> 25.0%
Cyclohexane extract	0-2000	2000-5000	5000-2.0%	2.0-10.0%	> 10.0%

Source: Kelly (1979)

The Kelly guidelines for contaminated soils provide a useful set of values for determining whether or not dredged material may be considered by a WRA to be contaminated, and if so, to what degree that contamination is present. The guidelines are widely known within WRAs and may be referred to for guidance on contamination during the assessment of licence applications. For example, the Greater Manchester WRA have established their own guidance criteria which set the limits of ranges for contamination levels for soils and other wastes (Greater Manchester WRA, 1994). The contamination classes reflect those suggested by Kelly, stating maximum values for each contamination class (i.e. A: uncontaminated, B: slightly contaminated, C: contaminated, D: heavily contaminated and E: unusually heavy contaminated). Where the same contaminants are considered by both sets of guidelines, Greater Manchester WRA's maximum values are typically the same as the upper range values suggested by Kelly; indicating how closely the Kelly guidance is followed.

The Kelly guidance does not include values for leachates or for contaminants such as PCBs, PAHs and chlorinated solvents. However overall, by comparing contaminant concentrations present in dredged material with the Kelly guidelines, a reasonable indication of the level of contamination may be established.

4.7.4 Interdepartmental Committee on the Redevelopment of Contaminated Land's (ICRCL) guidance on the redevelopment of contaminated land (ICRCL, 1987)

The ICRCL approach to the investigation and classification of a contaminated site follows a systematic method of assessment. Firstly the hazards (i.e. contamination) which are likely to affect land use should be identified. Then a site investigation for those contaminants which could give rise to such hazards should be undertaken. Lastly, an assessment of the level of contamination should be used for decision-making in terms of the site's end use. The ICRCL system is typically used for redevelopment of land with *in situ* contamination, and is not specifically established for identifying dredging disposal options. However, the ICRCL system may be appropriate for guidance on the disposal of dredged material to land where an end use (e.g. beneficial use or exempt use for agriculture or ecological improvement) may be planned. For example, if a use for the dredged material has been identified which includes the redevelopment of land for domestic gardens or allotments, then the ICRCL guidance concentrations may be referred to for an indication of the acceptance or rejection of the material in terms of contamination. Similarly, the ICRCL guidance concentrations may be referred to for guidance on contamination when seeking an exempt end use, for example a benefit to agriculture for "*any uses where plants are to be grown*" or for ecological improvement for "*parks*", "*open space*" or "*landscaped areas*" (ICRCL, 1987).

The trigger values for particular end uses have been established to aid the quantification of risks associated with contamination. Two sets of contaminant concentrations, known as threshold and action trigger values (see Table 4.7.4) are used to define the degree of contamination. Depending upon the degree of contamination, the contamination can be classified within one of three zones (see Box 4.7.4).

Box 4.7.4   ICRCL zones of contamination

- Contaminants are present in concentrations below threshold values, reflecting a low importance of any hazard. Treat as uncontaminated. No action required as risk no greater than is normally accepted.

- Contaminants are present in concentrations greater than threshold values but below action values. Significance of risk depends on intended use and form of development (i.e. assess importance of hazard). Use professional judgement to decide whether action is required.

- Contaminants are present in concentrations greater than action values. Risk is unacceptable and action is required as the contamination is hazardous. Treat as contaminated.

Source: ICRCL (1987)

In general, the ICRCL values can be used to indicate whether a disposal option and the proposed land reclamation end use for the dredged material (e.g. landscaping) will be acceptable in terms of soil contamination. However, it should be noted that the ICRCL guidance does not include trigger values for contaminants such as PCBs, pesticides, oils and other chlorinated hydrocarbons. If present in significantly high concentrations, these contaminants may prohibit the use of dredged material for various reclamation end uses irrespective of the ICRCL guidance. In addition, the ICRCL values provide no guidance on contaminant leaching. Overall, the ICRCL values are well known and recognised by regulators, and will provide a certain degree of guidance for planning dredged material disposal in terms of end use.

Table 4.7.4   ICRCL tentative trigger concentrations for selected contaminants

CONDITIONS

1. This table is invalid if reproduced without the conditions and footnotes.
2. All values are for concentrations determined on "spot" samples based on an adequate site investigation carried out prior to development. They do not apply to analysis of averaged, bulked or composited samples, nor to sites which have already been developed. All proposed values are tentative.
3. The lower values in Group A are similar to the limits for metal content of sewage sludge applied to agricultural land. The values in Group B are those above which phytoxicity is possible.
4. If all sample values are below the threshold concentrations then the site may be regarded as uncontaminated as far as the hazards from these contaminants are concerned and development may proceed. Above these concentrations, remedial action may be needed, especially if the contamination is still continuing. Above the action concentration, remedial action will be required or the form of development changed.

Contaminants	Planned uses	Trigger concentrations (mg/kg air-dried soil)	
		Threshold	Action
Group A: Contaminants which may pose hazards to health			
Arsenic	Domestic gardens, allotments	10	*
	Parks, playing fields, open space	40	*
Cadmium	Domestic gardens, allotments	3	*
	Parks, playing fields, open space	15	*
Chromium (hexavalent) (1)	Domestic gardens, allotments	25	*
	Parks, playing fields, open space		
Chromium (total)	Domestic gardens, allotments	600	*
	Parks, playing fields, open space	1000	*
Lead	Domestic gardens, allotments	500	*
	Parks, playing fields, open space	2000	*
Mercury	Domestic gardens, allotments	1	*
	Parks, playing fields, open space	20	*
Selenium	Domestic gardens, allotments	3	*
	Parks, playing fields, open space	6	*
Group B: Contaminants which are phytotoxic but not normally hazards to health			
Boron (water-soluble) (3)	Any uses where plants are to be grown (2, 6)	3	*
Copper (4, 5)	Any uses where plants are to be grown (2, 6)	130	*
Nickel (4, 5)	Any uses where plants are to be grown (2, 6)	70	*
Zinc (4, 5)	Any uses where plants are to be grown (2, 6)	300	*

NOTES:

* Action concentrations will be specified in the next edition of ICRCL 59/83.
1. Soluble hexavalent chromium extracted by 0.1M HCl at 37°C; solution adjusted to pH 1.0 if alkaline substances present.
2. The soil pH value is assumed to be about 6.5 and should be maintained at this value. If the pH falls, the toxic effects and the uptake of these elements will be increased.
3. Determined by standard ADAS method (soluble in hot water).
4. Total concentration (extractable by $HNO_3/HClO_4$).
5. The phytotoxic effects of copper, nickel and zinc may be additive. The trigger values given here are those applicable to the "worst-case"; phytotoxic effects may occur at these concentrations in acid, sandy soils. In neutral or alkaline soils phytotoxic effects are unlikely at these concentrations.
6. Grass is more resistant to phytotoxic effects than are most other plants and its growth may not be adversely affected at these concentrations.

Table 4.7.4   ICRCL tentative trigger concentrations for selected contaminants (continued)

CONDITIONS

1. This table is invalid if reproduced without the conditions and footnotes.
2. All values are for concentrations determined on "spot" samples based on an adequate site investigation carried out prior to development. They do not apply to analysis of averaged, bulked or composited samples, nor to sites which have already been developed.
3. Many of these values are preliminary and will require regular updating. They should not be applied without reference to the current edition of the report "Problems Arising from the Development of Gas Works and Similar Sites".
4. If all sample values are below the threshold concentrations then the site may be regarded as uncontaminated as far as the hazards from these contaminants are concerned, and development may proceed. Above these concentrations, remedial action may be needed, especially if the contamination is still continuing. Above the action concentrations, remedial action will be required or the form of development changed.

Contaminants	Proposed uses	Trigger concentrations (mg/kg air-dried soil)	
		Threshold	Action
Polyaromatic hydrocarbons (1, 2)	Domestic gardens, allotments, play areas.	50	500
	Landscaped areas, buildings, hard cover.	100	10,000
Phenols	Domestic gardens, allotments.	5	200
	Landscaped areas, buildings, hard cover.	5	1000
Free cyanide	Domestic gardens, allotments, landscaped areas.	25	500
	Buildings, hard cover.	100	500
Complex cyanides	Domestic gardens, allotments.	250	1000
	Landscaped areas.	250	5000
	Buildings, hard cover.	250	NL
Thiocyanate (2)	All proposed uses.	50	NL
Sulphate	Domestic gardens, allotments, landscaped areas.	2000	10,000
	Buildings (3).	2000 (3)	50,000 (3)
	Hard cover.	2000	NL
Sulphide	All proposed uses.	250	1000
Sulphur	All proposed uses.	5000	20,000
Acidity (pH less than)	Domestic gardens, allotments, landscaped areas.	pH5	pH3
	Buildings, hard cover.	NL	NL

NOTES:

NL No limit set as the contaminant does not pose a particular hazard for this use.
1. Used here as a marker for coal tar, for analytical reasons. See "Problems Arising from the Redevelopment of Gasworks and Similar Sites" Annex A1.
2. See "Problems Arising from the Redevelopment of Gasworks and Similar Sites" for details of analytical methods.
3. See also BRE Digest 250: Concrete in sulphate-bearing soils and groundwater.

Source: ICRCL (1987)

4.7.5 British Waterways Classification System for sediment

In 1992 British Waterways (BW) undertook their National Sediment Sampling Scheme to produce a "*national overview of dredging quality*" for their canals and navigations. This initiative was driven by UK Regulations under which all dredged material has to be considered as controlled waste as defined in the Environmental Protection Act 1990: Controlled Waste Regulations 1992 (SI 1992 No. 558). These Regulations stipulated that an accurate description of the waste is required if the waste is to be used, stored, handled or deposited. Additionally the Control of Pollution Act 1974: Collection and Disposal of Waste Regulations 1988 (SI 1988 No. 819) requires dredging to be disposed of at a licensed site where not exempt. In order to plan future dredging and disposal operations British Waterways surveyed channel sediments to quantify the concentrations of contaminating substances and to set up the Classification System for Sediments to represent these data qualitatively.

The classification system (see Table 4.7.5), updated to cover the Waste Management Licensing Regulations, was devised by integrating chemical quality criteria from three established sources in order to develop in-house guidance on the suitability of dredged material for specific uses.

Table 4.7.5 The British Waterways Classification System for Sediments

Class	Criteria
0	Not sampled - either due to: lack of sediment area inaccessible at time of survey (e.g. sample in tunnel).
A	Complies with Agricultural Use limits at pH >5* and does not exceed ICRCL thresholds. In addition does not have a Kelly rating greater than 0.3.
B	Complies with ICRCL thresholds with the exception of zinc, nickel, copper and boron. Threshold values for these phytotoxins are derived from ICRCL 70/90. Does not necessarily comply with thresholds for phenol and sulphide as ICRCL recommended methodology is inappropriate for organic-rich wet sediments. No single parameter in range of "unusually heavily contaminated" as defined by Kelly (1979). Does not have an overall Kelly rating greater than 1.5.
C	Exceeds ICRCL threshold for parameter other than zinc, nickel, copper, boron, phenol and sulphide. Has a Kelly rating greater than 1.5 and/or has one or more parameter in range of "unusually heavily contaminated" as defined by Kelly (1979).

* Lower pH values lead to greater mobility of metals

The British Waterways Classification System is based on the following:

- The most stringent classification limits in the system are based on the DoE's *Code of Practice for Agricultural Use of Sewage Sludge* (DoE, 1989b). This document establishes the maximum concentrations of certain toxic substances (metals and metalloids) permissible in soils used for agriculture where sewage sludge is disposed to this land. By implication, the BW classification considers these concentrations to be acceptable for agricultural land use when disposing of dredged material. BW also uses such concentrations to represent the "*most demanding classification limit*".

- The second source of guidance values is the ICRCL's threshold levels for land reclamation for use as open space and playing fields, and areas not used for growth of foodstuffs (ICRCL, 1987) (see Section 4.7.4). BW have adopted a precautionary approach, taking the threshold levels for parameters other than phytotoxins as the limits for Class B. Most of the substances listed under ICRCL criteria are metals and metalloids, but the criteria also include concentrations for PAHs and phenols.

- The third and final source of guidance used, covering a wider range of chemical substances, is based on the rating system devised by Kelly (1979) and the subsequent guidance developed by the former Greater London Council Waste Disposal Authority (see Section 4.7.3). The four contaminant concentration ranges using the Kelly values (calculated using an exponential formula), 0, 1, 2, and >2, approximate to the classifications of uncontaminated, contaminated (including Kelly's slightly contaminated category), heavily contaminated, and unusually heavily contaminated. Parameters not included in Kelly have been derived from the Dutch standards (e.g. tin) or by BW scientists (e.g. thallium and tungsten).

The range of substances included in the BW Classification System were based on previous analytical requirements of WRAs. Substances such as pesticides and mineral oils are not included.

Box 4.7.5 describes how the classification has been interpreted by BW to guide disposal options for dredged material.

Once the dredged material has been tested, the analytical results for a range of contaminants can be compared to the limits of the agricultural use, ICRCL and Kelly rating systems.

Box 4.7.5   Potential disposal options through classification

Class	Disposal Characteristics
A	Sediment likely to be suitable for disposal to agricultural land or adjacent land.
B	Sediment where concentrations of contamination are such that the sediment is likely to be suitable for disposal under exemption from the Waste Management Licensing Regulations 1994.
C	Sediment where concentrations of contamination are such that the sediment is unlikely to be suitable for disposal under exemption from the Waste Management Licensing Regulations 1994. Further investigation may be required.

The identification of a class for the dredgings will provide information which may be useful in the process of determining potential disposal options. This is unique to the BW system and is its major advantage. On the other hand, users should be aware that it can only be used for guidance, as WRA interpretation of the Waste Management Licensing Regulations may place different and more stringent emphases on certain aspects of disposal. This is particularly important when considering the deposition of dredged material to a proposed disposal site. For example, the system does not address the specific exemption requirements to demonstrate a benefit to agriculture or ecological improvement.

The system is derived for guidance and inclusion in one class does not preclude disposal in a different manner. In certain areas, for example, the deposit site characteristics may allow deposit as if the material were in a lesser class e.g. Class C material deposited under exemption as if it was Class B. The NRA are developing guidance on acceptable leachable values for contaminated material to be deposited in specified catchments and this may allow further flexibility in the deposit of material under exemption.

4.7.6 Classification guidance: conclusions

The differences between the classification schemes is accepted and acknowledged by the regulatory authorities. NAWRO is working towards closer co-operation and common interpretations between regulatory authorities. The formation of the NAWRO technical subgroup should facilitate this dialogue, as should the formation of the Environment Agency.

The conclusions on classification guidance are:

- In the UK there is no legally established scheme for classifying dredged material in terms of levels of contamination and potential impacts on the environment.

- Characterising the dredged material and classifying its degree of contamination could be an important element in the evaluation of an exemption registration for some exemptions (as well as disposal licences), and will, in some cases, be fundamental to the decision-making process of the WRA.

- The classification schemes do not represent legally binding, formal comparisons by which dredged material disposal options can be clearly evaluated.

- The lists of substances in each scheme are not necessarily appropriate for testing as part of a disposal application. Discussion between an operator and a WRA will identify for which substances data are required.

- The classification schemes help to identify the potential risk or impact; however, they do not provide guidance on how contamination may or may not affect the disposal site environment.

- Although classification schemes can be used to quantitatively evaluate the significance of any contamination present in a dredged material, their (scientific) use beyond the point of guidance is limited. This is particularly pertinent for large-scale disposal proposals for heavily contaminated dredged material. The crucial consideration to be recognised is that, for each dredged material and for each disposal site, the interactive (e.g. geochemical) changes between the sediment, soil and water media will be different.

- For assessing potential environmental and health risks, a quantitative classification scheme may be considered too rigid. It is therefore important to allow for a certain degree of flexibility when applying these classifications to disposal options, rather than interpreting their definitions as fixed and unmovable.

- All the classification schemes are useful in their own right, but one scheme may be more appropriate than the other two for guidance in certain situations. All these schemes should be used to give the best overall guidance on contamination.

5 Environmental issues

5.1 INTRODUCTION

5.1.1 Why consider the environment?

The dredging, handling, sorting, storing and disposal of dredged sediment all have the potential to cause detrimental environmental impacts. The potential impacts will be site specific. Conversely, appropriate disposal can have beneficial environmental impacts, such as the creation of new habitats.

Under the Waste Management Licensing Regulations, applications to a WRA for a disposal site licence will be carefully reviewed to ensure that no significant environmental effects are likely to occur. To that end, the WRAs and other competent authorities will lay down various conditions and monitoring requirements (e.g. in Waste Management Paper No. 26A (DoE, 1994c) on monitoring of licensed landfill sites).

Environmental considerations will therefore play an important role in determining the acceptability of different options for the use, storage, re-use or disposal of dredged material. Consideration of environmental impacts may result from legal requirements in the Regulations, the requirements of other statutory agencies who will be consulted by the WRA, guidance provided by non-statutory consultees, or simply best environmental practice.

In addition, dredgings disposal carried out for or by authorities that have a Duty to further conservation (e.g. NRA and BW) must consider environmental issues in accordance with their environmental codes of practice.

5.1.2 Supporting Information

In order to confirm an exemption or grant a licence, the WRA will expect to see evidence supporting the applicant's conclusions on environmental benefits (e.g. benefits to agriculture or ecological improvement) and/or on the minimisation of potential adverse environmental impacts. The WRA will similarly expect the applicant to have dealt with the concerns of other statutory consultees (e.g. NRA, English Nature, Health and Safety Executive). This may, in some cases, involve the applicant in undertaking discussions, evaluating options and producing a report.

A full, formal Environmental Assessment (see Section 5.2.2) is rarely required, but one can be requested, in particular if planning permission for a new licensed disposal site is being assessed.

5.1.3 Consultation

Whether undertaking a formal Environmental Assessment or a less formal investigation, consultation with the WRA, the NRA and other statutory consultees, and other interested parties, at the earliest possible opportunity will be essential if environmental concerns are to be identified and addressed in the most cost-effective manner. Such consultation, particularly during the scoping exercise (see Table 5.2.3), will help to focus any investigations on the points of particular concern for the materials and/or the receiving site(s) under consideration. Ongoing consultation will then be essential to ensure that any problems identified are subsequently resolved to the satisfaction of the responsible agency.

A list of some of those agencies who might have interests in a particular site (both statutory and non-statutory), and/or who might be able to supply relevant data, is provided in Table 2.3.2. Those consulted should be able to advise the applicant on their particular concerns and on the legal and/or regulatory framework within which they expect the applicant to operate. This latter point is important because there may be other relevant legislation in addition to the Waste Management Licensing Regulations.

Finally, consultation is especially important while the Waste Management Licensing Regulations are still in their infancy as both parties will have the opportunity to develop an understanding of the other's requirements, while working to identify a mutually acceptable solution.

5.2 PRINCIPLES OF ENVIRONMENTAL APPRAISAL

5.2.1 Background

This chapter uses "environmental appraisal" as a general term to cover all types of analysis or evaluation for the resolution of environmental issues. The value of environmental appraisal as an evaluation procedure is described in Chapter 2 of the CIRIA publication "Environmental Assessment: a guide to the identification, evaluation and mitigation of environmental issues in construction schemes" (CIRIA, 1994). Chapter 9 of the same document deals with waste management.

Sections 5.2.2 onwards of this guidance document deal specifically with those environmental aspects of dredgings disposal which differ in some way from conventional waste management issues, for example because of the nature of the materials being handled or because of the specific requirements of the Waste Management Licensing Regulations.

5.2.2 Need for formal Environmental Assessment

Where the WRA (or, if planning permission is required, the planning authority) feel that the environmental effects of a proposed waste disposal operation may be "significant", an Environmental Statement (the report resulting from the formal Environmental Assessment process; see Figure 5.2.2) can be required to accompany the application. English Nature, the NRA, or other statutory consultees can similarly request, via the licensing authority, that an Environmental Assessment be undertaken. "Significance" can be a function of, *inter alia,* the sensitivity of the site, its size, etc.

If a formal Environmental Assessment is required for a proposed storage, disposal or use option, there are particular requirements as set out in the Statutory Instruments implementing EC Directive 337/85/EEC (see Section 3.9.5). Environmental Statements prepared under this legislation should contain certain specified information with regard to the characteristics of the project, the study area, the effects of the project on the study area and appropriate mitigating measures.

Depending on the amount of data which needs to be collected, an Environmental Assessment will generally take 3-8 months to complete. If seasonal biological data has to be collected, the assessment can take up to 18 months.

5.2.3 The environmental appraisal process

Irrespective of whether or not a formal Environmental Assessment is required, the environmental appraisal process (see Table 5.2.3 and Figure 2.7.1) provides a useful framework for the evaluation of disposal options.

The Environmental Assessment process shown in Figure 5.2.2 provides a structured approach which enables a dredging operator to identify all the relevant environmental aspects involved in a disposal option and prepare his disposal application to a WRA.

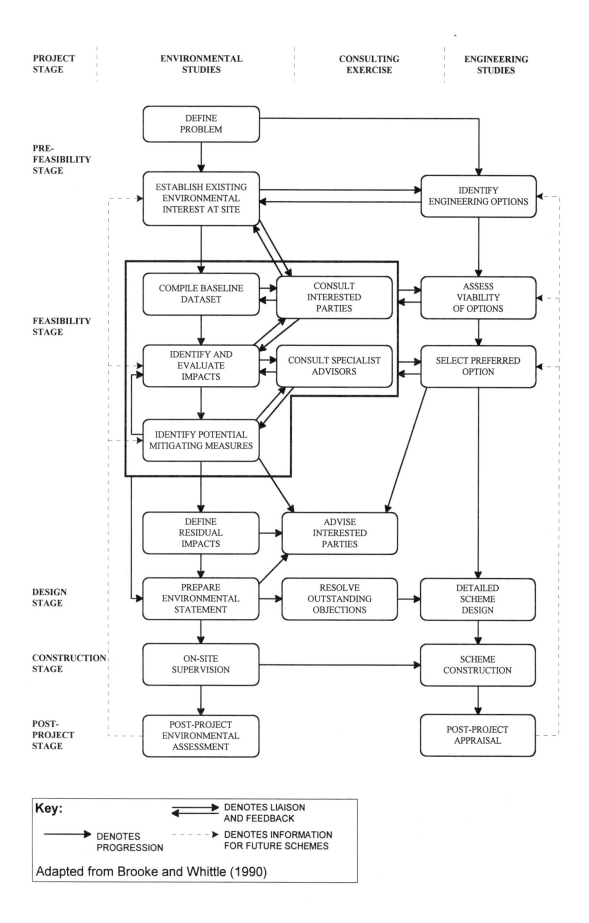

Figure 5.2.2 *The Environmental Assessment process*

Table 5.2.3   Environmental appraisal activities

Activity	Description
Scoping	Determines the key environmental issues at a particular site or for a particular material; comprises a preliminary data collection exercise, consultation and review of potentially significant impacts; is particularly important in ensuring investigations are carefully targeted and hence cost-effective.
Data collection	See Section 5.3.2.
Consultation	See Section 5.1.3.
Impact identification Impact prediction and evaluation	Based on *inter alia*, consultation, data collection, literature search and review, etc. Can include qualitative and quantitative assessments, risk assessment, mathematical or computer modelling, etc. Significance is determined against relevant standards, thresholds, etc. See also Section 5.3.1.
Identification and appraisal of mitigation measures	See Section 5.4.
Identification of environmental enhancement measures	May be opportunity to provide environmental enhancement (e.g. landscaping, planting, amenity provision) at little or no extra cost.

5.2.4 Resolving environmental issues

Operators should establish, at the earliest possible stage, whether or not a formal Environmental Assessment will be required. If a full Environmental Assessment is required, or if specific environmental problems need resolving, the operator may undertake such work internally, or specialist environmental consultants might be employed. One advantage of the latter, in addition to experience in resolving environmental issues, is that the assessment will be seen to be independent.

In order to progress his/her application, the operator should identify the key environmental issues through consultation and checklists based on the following sections, and ascertain the type and detail of data required.

5.3 ENVIRONMENTAL DATA COLLECTION AND IMPACT EVALUATION

5.3.1 Possible environmental impacts

Impacts on the environment can be either positive or negative. Where negative impacts are identified, mitigation measures will be required. Box 5.3.1 sets out examples of possible environmental impacts of dredged sediment disposal (storage, re-use, etc.) to land. It should be stressed, however, that this list is not definitive and in many cases only a few of the impacts will be of concern, indeed dredging disposal sites normally pose less risk to the environment than the majority of commercial sites. In addition to positive or negative, it should also be noted that impacts can be:

- short- or long-term
- reversible or irreversible
- local or strategic
- direct or indirect
- cumulative.

Box 5.3.1   Examples of potential environmental impacts

Characteristic	Potential impact[1]
Water quality	Contamination of watercourses via surface runoff/flow through soil profile. Contamination of groundwater by leaching.
Soil quality	Change to soil physical characteristics. Bio-accumulation of contaminants in soil. Effect of contaminants on soil biota.
Air quality	Windblown dust from disposal site. Vehicle emissions from HGVs used to transport material. Landfill gas emissions.
Ecosystem	Habitat destruction. Ecological improvement (e.g. habitat creation). Bio-accumulation of contaminants in plants and animals. Special loss.
Land use	Change of use. Restrictions on or opportunities for future use(s).
Landscape	Degradation of landscape. Landscape improvements. Damage to archaeological features.
Human activities	Disruption to footpaths, towpaths, etc. Temporary disruption due to transportation or construction activities. Beneficial use of materials (e.g. to improve flood defence).
Human health	Effects of airborne pollutants. Effects of contaminated water. Effects of eating contaminated foodstuffs. Disturbance due to noise.

Notes: (1) Excludes indirect impacts.

5.3.2 Data collection

In order to assess the potential impacts of the storage, re-use or disposal of dredged material on a receiving site, it is important to develop an understanding of the interactions which may lead to such impacts through data collection and review. The environmental characteristics of the receiving site which might be affected in some way include, amongst others:

- ground and surface water (e.g. proximity of surface water or groundwater resources and the use made of them; hydrology; hydrogeology; etc.)
- soil quality (soil chemistry; geology; etc.)
- air quality
- ecosystem/habitats (e.g. proximity of designated sites of nature conservation importance or presence of scarce or protected species)
- land use (e.g. current land uses at and adjacent to the proposed receiving site including residential areas, infrastructure, etc.)
- landscape (including protective designations)
- human activities (e.g. recreation or amenity)
- human health (population characteristics, etc.).

The consultation process will help the collation of data currently held by interested parties. Other sources of information include Ordnance Survey maps, local libraries or universities. If no information exists, particularly on the characteristics of the receiving site, site surveys and/or primary data collection may be necessary. In this case the representative agency (e.g. the WRA, NRA, English Nature, etc.) should be consulted to establish precisely what information needs to be collected to satisfy their concerns, and why. Such consultation is likely to be particularly pertinent when notifying a proposed exempt disposal activity for the first time (see Sections 5.5 and 5.6).

Once data has been collated, a thorough review by experts, possibly combined with consultation, will need to be undertaken in order to certify potential impacts requiring evaluation.

5.3.3 Impact evaluation

Potential impacts can be evaluated in a number of ways. The evaluation of sediment quality is discussed in detail in Chapter 4. Many other impacts can be resolved simply through discussion and consultation with interested parties. In some cases, a more technical evaluation will be required (e.g. NRA's leaching test as discussed in Section 4.4.4).

Two particular further techniques may be appropriate in certain circumstances. These are only briefly described below as their application is very site specific and it is likely that specialist assistance would need to be sought.

* Risk assessment

In assessing potential environmental impacts, it is sometimes necessary to evaluate the likelihood of a particular event or chain of events. Risk assessment is applied systematically to assess the level of risk associated with a given activity and identifies ways by which risks can be reduced. Assessments vary in complexity according to need but all follow the same systematic approach which involves finding answers to the following questions:

- what can go wrong (i.e. what risks does this material pose)?
- how likely is it to go wrong?
- what would happen if it did go wrong (i.e. what is the nature of the risk)?
- what are the associated levels of risk?
- are the risks acceptable?
- can the risks be reduced?

Answering some of these questions will require a thorough analytical appraisal of the material and an understanding of possible interactions with the receiving site.

The first and most important step in any risk assessment is to define the risk of concern. Some of the risks associated with the disposal of dredged sediments are already covered by existing legislation (e.g. existing health and safety legislation requires that a simple risk assessment is carried out to assess the risks to operators associated with the handling of materials, etc.) and some are explicitly dealt with as part of engineering design requirements (e.g. the risk of a lagoon bund failure is minimised by designing the bund to an appropriate factor of safety). *Environmental Assessment. A guide to the procedures* (DoE, 1989a) states that where health and safety is a major issue, a full hazard and risk assessment may be required as part of a separate study. Similarly, consultations with WRAs and the NRA have revealed that both competent authorities will require proof that the disposal of potentially contaminated material does not pose an undue risk before licences (or exemptions) are granted. In summary, the potential risks, which will vary according to the nature of the disposal site and the nature of the disposed materials, include:

- risks to people associated with possible direct contact with any highly contaminated materials
- risks to ground and surface water associated with the leaching of toxic substances
- risks to surface waters (and the life they support) associated with the spreading of material on land and with bankside disposal
- risks to people (and to nearby property) associated with the generation of methane gas at a landfill site
- risks of phytotoxic effects due to presence of certain heavy metals
- risks to people (and the operations of the lagoon) associated with a bund failure.

Clearly, the level (and nature) of risk associated with each disposal operation will be different. In some cases the risks will be insignificant, while in others the risks may be high enough to require that action is taken to mitigate them or, in the extreme, to prevent the disposal operation from taking place.

- Modelling

Modelling can play an important role in environmental evaluation, facilitating a detailed investigation of the possible interactions between environmental variables and the associated impacts. Models are designed to simulate the responses of environmental systems under different conditions. There are three main potentially relevant approaches in this respect: statistical forecasting techniques based on trends, etc.; modelling using formulae (mathematical); and computer modelling. Modelling may need to be used where the task of data collection and interpretation is insufficient to predict interactions between parameters or waste disposal impacts on the receiving site. Modelling and forecasting are only as good, however, as the data that goes into them. The results will, therefore, need to be interpreted with an appropriate element of caution.

5.4 MITIGATION MEASURES AND MONITORING REQUIREMENTS

5.4.1 Need for mitigation

In cases where potentially significant adverse environmental impacts are identified, it will be necessary to review options for mitigation. Possible opportunities for cost-effective environmental enhancement might also be explored. The objective of mitigation is to reduce the potential impact to an acceptable (i.e. insignificant) level or to eliminate the impact altogether.

Assuming the environmental investigations have been initiated sufficiently early in the option selection process, it will generally be possible to identify technically viable and cost-effective mitigation measures. In some cases, mitigation can be achieved at little or no extra cost. In others, costs might be incurred, and these costs will therefore need to be balanced against a more expensive but environmentally acceptable option.

5.4.2 Mitigation options

Mitigating possible contamination or human health effects is likely to be of greatest concern to the WRA and NRA. Mitigation may, however, also be necessary to reduce some of the other possible impacts. Table 5.4.2 sets out some examples of mitigation across a range of potential impacts.

Table 5.4.2 Examples of mitigation measures

Potential impact	Possible mitigation measures
Ground or surface water quality deterioration	Appropriate disposal options or treatment prior to disposal; lining; capping; leachate control.
Visual intrusion	Landscaping (e.g. planting); and reduction in height or area.
Habitat destruction	Selective clearance or placing; planting/habitat creation nearby* or at end of use as disposal site.
Air quality deterioration	Use bulk haul transfer.
Disruption/loss of access	Provide temporary access; divert footpath.
Destruction due to bund failure	Carefully controlled design/construction.

* Indirect mitigation measures

5.4.3 Mitigation to prevent the spread of contamination

Mitigation measures and the implementation of a case-specific monitoring programme are likely to be required when it has been established that a material may cause contamination.

In the context of the possible release of contaminants, mitigation measures are generally designed to prevent the release or dispersion of contaminants or other agents (e.g. suspended solids) which could have an environmental impact. Such measures will be selected on a case-specific basis according to the sediment's and the disposal site's properties. In some cases a combination of mitigation measures may be needed.

Mitigation measures for contaminated sediment disposal can be directed either towards preventing the release of any effluent from the disposal site (5.4.4) or towards treating the released effluent to an acceptable standard before it enters surface water or groundwater (5.4.5).

5.4.4 Prevention of effluent release

The release of effluent from contaminated dredged sediment can be prevented either by selecting a disposal site with natural mitigating properties or by applying mitigating measures such as disposal site lining. Proper site selection can ensure minimal surface runoff or release of contaminants by flooding. A site with natural clay lining may also help to prevent the dispersion of contaminated leachates.

Avoiding sandy areas or aquifer recharge areas when selecting an onland disposal site can significantly reduce the possibility of contamination of groundwater. NRA is highly unlikely to permit the disposal of contaminated dredged material in aquifer recharge areas. The need to select an onland disposal site with natural mitigation properties against runoff or leaching where necessary cannot be sufficiently emphasised.

If a site with natural mitigation properties is not available, the following options to prevent or control the release of contaminated effluent should be considered:

* lining the disposal site using either natural (e.g. clay) or synthetic liners (e.g. high-density polyethylene (HDPE) or polymerised vinyl chloride)
* capping the disposal site; capping also reduces the possibility of wind-blown contaminants
* leachate collection for treatment
* hydrogeological controls used to prevent the dispersion and escape of leachates (a technique used in the Netherlands at the Slufter disposal site (Csiti, 1993)).

The effectiveness of natural liners is case-specific. Depending on the end use of a disposal site, the capping layer may or may not be of a specific permeability. Landscaping a capped disposal site would for instance require a permeable capping layer, to allow a root system to develop.

Should there be a need to use any of the above-mentioned mitigation techniques, specialist advice should be sought.

5.4.5 Treatment of released effluent

The oldest and simplest method of leachate collection and treatment is based on the principle of sedimentation. This is achieved by disposing of dredged sediment in a series of lagoons, and allowing the water to overflow naturally from lagoon to lagoon while the suspended solids settle out. The final effluent can usually be returned to the nearest surface watercourse with very low levels of suspended solids in accordance with the NRAs relevant discharge consent. Suspended solids can also be removed by chemical clarification and filtration.

The removal of dissolved metals and organic contaminants can be achieved in a number of ways by applying either physico-chemical or biological agents or a combination of the two (see Section 6.4). One or more of the following techniques can be used: chemical precipitation, filtration, carbon adsorption, ozonisation, distillation, electrodialysis, reverse osmosis and ion exchange. All these techniques, particularly the latter four, may be expensive, and not all of them have been used extensively. The use of any of these techniques must be based on specialist advice.

Finally, it should be noted that if the effluent released from a disposal site contains high biochemical oxygen demand (BOD), ammonia or sulphate concentrations, it should be routed via a sewage treatment works before release into the environment.

5.4.6 Monitoring requirements

Monitoring requirements comprise statutory licensing requirements for operation monitoring (Section 5.4.7) and/or mitigation measures identified by an Environmental Assessment covering both site construction and operational phases. Monitoring is inherent to the Waste Management Licensing system.

The Environmental Assessment process may determine that the disposal of dredged material at a licensed waste disposal site could generate adverse environmental impacts, and so monitoring is required. Examples of possible impacts during the construction phase include landscape, land use, local community and nature conservation.

Dredging disposal facilities should not be treated as commercial landfills and may need less monitoring and supervision than a commercial site.

Examples of waste management impacts which may require some form of monitoring are listed in Table 5.4.6. In some circumstances (e.g. leachate control, gas generation), however, site monitoring will be a statutory requirement of the licensing procedure.

Table 5.4.6   Environmental issues where monitoring may be required to facilitate appropriate mitigation

Impact[1]	Possible cause	Effects
Acute/chronic groundwater pollution	Seepage/dispersal from site (e.g. liner damaged)	Aquifer contamination
Acute/chronic surface water pollution	Leachate contaminating clean discharge; seepage	Fish kills; reduced water quality
Particulates	Dust from earthworks and landfill operation	Nuisance; visual quality
Destruction of habitat	Site preparation; clearance; operation	Loss of habitats, species
Health and safety of operators	Noxious gases; skin contact with contaminants	Health risk

(1)   Further impacts are presented and discussed in detail in Section 9.4 of CIRIA's Environmental Assessment guidance document (CIRIA, 1994).

5.4.7 Monitoring licensed disposal sites

Disposal sites licensed to accept waste which requires containment are subject to monitoring requirements to prevent adverse environmental impacts from occurring (e.g. water contamination or combustible gas generation) and ensure that confinement measures (e.g. liners) are working.

Waste Management Papers Nos. 4, 26, 26A, 26B and 27 provide guidance on the monitoring required to gain and maintain a licence and subsequently obtain a Certificate of Completion. These Papers set out guidance on:

- the number of sampling points for surface water (where necessary), groundwater (where necessary), leachates and landfill gas

- determinants and monitoring frequencies for surface waters, groundwaters and background gas levels to monitor the baseline situation at the site preparation phase

- determinants and monitoring frequencies for surface waters, groundwaters, leachates and landfill gas at the site operation phase, the post-closure phase, and the site completion phase.

Table C3 (see Table 5.4.7) in Waste Management Paper No. 4 (DoE, 1994b) gives a broad indication of the monitoring requirements during the operation of a licensed landfill disposal site. Operators should bear in mind, however, that monitoring requirements are case-specific and therefore specialist advice should be sought.

It is important for early consultation to take place between an applicant (future licensee) and a WRA to establish the likely monitoring requirements and associated legal obligations established as part of the Waste Management Licensing system. The WRA will also consult the NRA or other relevant water pollution control authority on such matters.

Monitoring is the responsibility of the licensee, but the WRA may verify the monitoring data by carrying out its own analyses and cross-checking records. Monitoring requirements for licensed disposal sites can sometimes prove costly in terms of finance, time and effort, particularly if the material is highly contaminated, or if disposal takes place in an environmentally sensitive location. All such matters must be taken into account before applying for a licence.

Other environmental characteristics which may require some monitoring at a licensed disposal site include controls on fires, loose waste, windblown waste, weeds, vermin and insects, dust, cleanliness of vehicles and the sheeting of waste loaded in open containers.

Table 5.4.7   Monitoring requirements during the operation of a licensed landfill disposal site

Requirement	Monitoring frequency	Determinants
Surface water (if necessary)	Monthly	pH, Temp, electrical conductivity (EC), dissolved oxygen (DO), ammoniacal nitrogen (NH_4-N), chloride (Cl), chemical oxygen demand (COD).
Groundwater (where necessary)	Monthly Quarterly (can be reduced to 6-monthly)	Water level, pH, EC, Temp, DO, NH_4-N, Cl. As above plus: sulphate (SO_4), alkalinity (Alk), total oxidised nitrogen (TON), total organic carbon (TOC), sodium (Na), potassium (K), calcium (Ca), magnesium (Mg), iron (Fe), manganese (Mn), cadmium (Cd), chromium (Cr), copper (Cu), nickel (Ni), lead (Pb), zinc (Zn).
Leachate at discharge points	Weekly Monthly (can be reduced to quarterly) Quarterly 6-monthly (can be reduced to annually)	Discharge volume pH, Temp, EC. As weekly plus: NH_4-N, Cl, biochemical oxygen demand (BOD), COD. As monthly plus: SO_4, Alk, TON, TOC, Na, K, Ca, Mg. As quarterly plus: Fe, Mn, Cd, Cr, Cu, Ni, Pb, Zn.
Leachate at monitoring points	Monthly Quarterly (can be reduced to annually) Annually	Leachate level, pH, Temp, EC. As monthly plus: Cl, NH_4-N, SO_4, Alk, COD, BOD, TON, TOC, Na, K, Ca, Mg. As quarterly plus: Fe, Mn, Cd, Cr, Cu, Ni, Pb, Zn.
Landfill gas	As Waste Management Paper No. 27 (1991)	Methane (CH_4), carbon dioxide (CO_2), oxygen (O_2), atmospheric pressure (AP), other meterological data (OMD), Temp.
Other parameters	Annually	Void utilisation, settlement.

Source: Table C3 Waste Management Paper No. 4 (DoE, 1994b)

5.5 USING DREDGED MATERIAL AS A BENEFIT TO AGRICULTURE

5.5.1 Introduction

Dredged material, by its nature, may have physical and/or chemical characteristics which, if it is mixed with a soil, can provide benefit to agriculture. This potential use of dredged material is recognised under the Waste Management Licensing Regulations, which provide for licence exemptions in certain circumstances where sediment disposal has a benefit to agriculture. The DoE's Waste Management Licensing guidance notes provide some indication of what may be regarded as a benefit to agriculture (see also Section 3.5.5).

Analysis of the sediment (see Section 4.4) should provide important information about any potential beneficial or adverse physical and chemical properties of a dredged material. Testing needs to cover the potentially toxic elements (PTEs) set out in *The Code of Practice for Agriculture Use of Sewage Sludge* (DoE, 1989b) as well as potentially beneficial properties such as particle size distribution, organic matter content, nutrients, pH, etc. Operators need to seek specialist advice, but in all cases they should take advice from NRA's document *Policy and Practice for the Protection of Groundwater* (NRA, 1992).

Potential benefits due to the application of dredged sediment may include physical and chemical benefits.

Physical benefits to soil (primarily soil texture and structure):

- water-holding capacity
- water and air movement
- workability of the soil
- increased nutrient availability
- reduced soil compaction/erosion
- reduce or avoid flooding by filling or levelling hollows.

Chemical benefits to soil:

- fertilisation with nutrients
- increased nutrient availability
- pH adjustment.

This remainder of Section 5.5 briefly discusses potential benefits to agriculture within the framework of licence exemptions and describes the physical and chemical properties of dredged material which can benefit soils.

5.5.2 Exemption

Consultation highlighted the WRAs' strong reliance on the DoE guidance notes in relation to this exemption. The relevant DoE guidance (DoE, 1994; 1995) paragraphs include 5.66 to 5.86. Particular importance is likely to be placed on paragraph 5.74 which states "*in order to keep within the terms of the exemption it will be essential to establish on the basis of properly qualified advice what application rate is appropriate for each waste material, each soil and each site*".

The WRAs are likely to require assessments on a case-by-case basis, including interpretation "*according to the type of wastes and the category of land involved*". The DoE guidance also indicates that there is no formally adopted set of waste and soil characteristics which can identify simply whether or not the disposal option provides a benefit to agriculture. Thus, the spreading of a dredged material on agricultural land will have to be evaluated on its individual merits.

Most WRAs are likely to require the applicant to produce "properly qualified advice" on whether an agricultural benefit can be achieved. Such advice may require the use of an agricultural specialist. For the exemption to be valid, the operator responsible for the disposal of the dredged material must present the WRA with relevant information for the area where the sediment is proposed to be applied. This is the responsibility of the "*establishment or undertaking carrying out the spreading, who is not necessarily the owner or occupier of the land*" (DoE, 1994; 1995).

5.5.3 Potential benefits of dredged material on agricultural land

There are several potential beneficial uses of dredged material for agricultural purposes. However, to be acceptable to the WRA, benefits may be limited to marginal soils on land already under agricultural use. Table 5.5.3 indicates the range of assessment parameters that may be required to demonstrate that the proposed spreading will constitute an agricultural benefit.

Table 5.5.3   Information on dredged material and the receiving site that may require consideration for decision-making on benefits to agriculture

Chemical characteristics	Physical characteristics	Site proposed for spreading operation
Nutrients	Texture	Soil association
Metals	Structure	Soil (land) classification
Salinity	Water content	Agricultural use
Liming properties/requirements	Workability	Existing farm management
	Compaction	

5.5.4 Physical characteristics

The physical properties of dredged material are likely to be fundamental to decision-making on the potential benefits (or adverse impacts) to the receiving agricultural environment. The texture of the sediment relative to the soil texture at the disposal site, for example, can lead to beneficial qualities for crop growth.

- Texture

A sediment's texture is basically its grain-size distribution. Texture has an important influence on the movement of water and air, and affects the overall workability of a soil. The grain size itself has an important function in terms of concentrating substances. A coarse-grained sandy sediment is likely to be low in organic matter, nutrients and toxic substances (such as heavy metals). Organic matter is also an important textural component and can affect the fertility of a soil.

A textural benefit to agriculture may be achieved by the mixing of a fine-textured dredged material (clay or silt) with a coarse-grained marginal soil (sand) to produce a loam soil. A loam texture can typically produce all-round beneficial soil properties such as water movement, workability and nutrient availability. Alternatively, sandy sediment may improve the structure, permeability and workability of heavy, impermeable clay (Spaine *et al*, 1978) (see Table 5.5.4).

When considering the use of dredged material for potential benefit to agriculture, the nature of the existing or proposed agricultural use also needs to be taken into account, in terms of potential physical damage to the soil. Guidance on these issues can be found in MAFF's *Code of Good Agricultural Practice for the Protection of Soil* (MAFF, 1993). In addition, specialist advice must be sought.

It is possible that extraneous objects such as rubble, refuse and other wastes could be found in dredged sediment. Such waste materials will need to be removed to ensure there is no risk associated with their presence (for example, large solid objects being caught in farm machinery, or glass being ingested by cattle).

Table 5.5.4   Examples of beneficial dredgings for soil texture and fertility

Receiving soil	Most beneficial dredgings	Comments
Coarse-grained sandy soil	Clays and silts	Creation of a loam soil to improve structure, workability and water holding capacity (Spaine *et al*, 1978).
Heavy, impermeable clay	Sands	Improvements to soil structure, workability, water and air movement.
Infertile sandy soil	Fine sediments with high organic matter content (c.20-30%)	Improve soil structure, workability, water holding and nutrient availability.

5.5.5   Chemical characteristics

- Nutrients

Plants require readily available supplies of nutrients to grow satisfactorily. Sediments from dredged waterways may incorporate a variety of substances, including nutrients from soil minerals, fertiliser runoff, etc. Sediment analyses can provide important data on available nutrients. Some metals are important trace nutrients for plants (e.g. iron, manganese, copper, zinc, molybdenum, boron); however, in high quantities these metals are phytotoxic (i.e. poisonous to plants). The nutrients of major interest in respect of agricultural improvement are nitrogen, phosphorus, the metals potassium, calcium and magnesium, and sulphur. Nutrients are most likely to be contained in fine- and medium-grained dredged material. Consultation has indicated, however, that a dredged sediment's value as a soil fertility improver may be limited.

- Metals

Significant quantities of metals (i.e. potentially toxic elements: PTEs) present a potential risk to crop growth and animals, including humans, through consumption and accumulation, biological activity in the soil, and/or soil and groundwater quality (MAFF, 1993). Any metals (including trace nutrients) present in the dredged material may therefore pose an environmental risk following spreading on land. The uptake of potentially toxic metals in significant quantities is mainly dependent on the form and the available concentration of the metals, and the plant species itself.

MAFF's primary concern regarding the disposal of dredged material to agricultural land would be soil contamination and its effects. In the UK there are restrictions on the levels of metals in soils to which sewage sludge (or dredged material) is applied. The *Code of Practice for Agricultural Use of Sewage Sludge* (DoE, 1989b) establishes maximum concentration for various metals (as PTEs) in soils based on different pH ranges. Chemical analyses of the dredgings will probably be required to prove no disbenefit to soils in this respect. Some testing of the soil itself may also be required to assess local background levels of naturally occurring metals, and/or whether sewage sludge has been applied to the land. The data presented in the DoE guidance can be used as a basic indicator to the levels of metals in dredged material which may be considered to have a potential adverse impact when spread on agricultural land.

The *Code of Good Agricultural Practice for the Protection of Soil* (MAFF, 1993) includes the sewage sludge metal levels as terms of reference for concentrations of potentially toxic elements in the soil after applications of material to agricultural land. With regard to protection of the soil, it is the farmer/landowner's responsibility to obtain proof from the operator that the application of dredged material would have no disbenefit to land use. This does not overrule any requirements of the WRA. Therefore, to demonstrate that dredged material spreading is beneficial, it will have to be shown that there are no adverse impacts.

• pH

The pH of dredged material is important in determining the beneficial use of dredged material for agriculture. A pH of 4 or below indicates very acidic sediment/soil conditions, which would support few plants, and would be unsuitable for agricultural purposes. A higher pH can also decrease the uptake of metals by plants or crops (Spaine *et al*, 1978). Table 5.5.5(a) demonstrates the optimum pH range for a variety of plant growth. If the sediment has a pH less than 6.5, it may require some application of lime before being applied to agricultural land, in order to be of benefit to agriculture. Liming could be used if the dredged material contained large concentrations of sulphur, as it can mitigate against the generation of acidic conditions.

Table 5.5.5(a)   Soil pH and agricultural condition

pH of soil	Conditions for plant growth
c.5.8-7.5	Optimum pH range for arable crops
c.5.2-7.5	Optimum pH range for grassland
c.3.8-8.2	pH range for natural and semi-natural habitats and rough grazing
7	Neutral conditions
<7	Increasing acidity
>7	Increasing alkalinity

Source: Information to produce this table has been taken from MAFF (1993).

• Salinity

Sediment dredged from within the tidal areas of estuaries, ports and harbours will have a degree of salinity. The concentration of the soluble salts can be estimated by measuring the electrical conductivity of a sediment water extract. Excess concentrations of soluble salts limit the availability of water to plants and restrict growth. Table 5.5.5(b) provides guidance on crop responses to various soil-water salinities.

Table 5.5.5(b)   USDA recommendations for plant growth on saline soils

Electrical conductivity (mmho/cm)	Effect on plant growth
<2	Typically negligible effects.
2 - 4	Yields of sensitive crops may be restricted.
4 - 8	Yields of most crops will be restricted.
8 - 16	Only salt-tolerant crops produce satisfactory yields.
>16	Only very salt-tolerant crops produce satisfactory yields.

Source: Information to produce this table has been taken from Spaine *et al* (1978)

5.5.6 Assessing benefits to agriculture

Spaine *et al* (1978) identify the following factors which need to be understood at a proposed disposal area: the properties of the receiving soil, the application depth of the dredged material, soil compaction, changes to erosion potential, changes to flooding and drainage regimes, and land preparation (i.e. tillage, cultivation and planting).

The three exemptions involving the spreading of dredged material on land where a benefit to agriculture has to be shown are discussed below.

• Spreading on agricultural land

Land already under agricultural production is likely to have some existing information available about its physical, chemical and biological characteristics. For example, testing for nutrient availability may have been carried out to identify fertiliser application requirements. It is unlikely that good quality agricultural land (e.g. MAFF classification grades 1, 2 and 3a) would be significantly improved by the spreading of dredged material to constitute a benefit. However, the application of dredged material to all good quality land should not be simply disregarded. For example, drought-prone grade 3a soils might be improved by clay.

The spreading of dredged material is more likely to be of benefit on currently marginal soils where the economic returns for crop production are relatively low (e.g. unproductive pastures, abandoned fields, fields requiring excessive irrigation, etc.) through the adding of dredged material to produce a loam.

• Spreading for land reclamation and improvement

If land that may contain some degree of contamination is to be reclaimed or improved, action should be taken to identify the existing concentrations of contaminating substances. This is particularly important if redeveloping former industrial sites for agricultural use (MAFF, 1993). Consultation with the WRAs during the preparation of this document has indicated that the ICRCL's *Guidance on the Assessment and Redevelopment of Contaminated Land* (ICRCL, 1987) should not be used as guidance for this purpose. Operators should therefore refer to the guidelines discussed in Section 5.5.5 referring to potentially toxic elements in soils for agriculture.

• Spreading on banks and towpaths

In many cases the option to dispose of dredged materials on banks and towpaths where dredging takes place will be exempt irrespective of the potential requirement to show a benefit to agriculture (or ecological improvement). However, where the exemption is to be justified on the basis of a benefit to agriculture, it is likely that such spreading will be limited due to the size, location and nature of banks and towpaths, as well as the actual use of such areas for agriculture (possibly with the exception of grazing). To a certain extent the importance of this exemption will vary with the definition of "bank". The Jones v River Mersey Board case would suggest that this could be a substantial area (see Section 3.7.2).

5.5.7 Summary of requirements for benefits for agricultural exemption

The series of investigations undertaken during the production of this guidance document has highlighted the following:

• Although simply raising or changing the level of the land by the bulk application of dredged material may not constitute a benefit to agriculture (see Section 3.5), raising the level of land could prevent flooding.

• It is likely to be easier to show that the spreading of dredged material represents a benefit to agriculture on marginal soils (e.g. MAFF land classification grades 3b, 4 and 5) rather than existing good quality farmland (MAFF grades 1, 2 and 3a).

• Possible physical benefits to the soil's texture (and structure) may include improved water and air movement, water-holding capacity, nutrient availability and workability. The relative mixture of sediment and soil's grain sizes (clays, silts and sands), and their organic matter content, will determine any textural benefits.

- Chemical benefits are likely to be more difficult to demonstrate. For example, the use of dredged material as a nutrient source may also be beneficial if a textural benefit is similarly achievable in terms of improving nutrient availability.

- Demonstrating a beneficial application of dredged material to soil may not just involve identifying benefits to agriculture. The spreading of sediment must not have any detrimental effects.

- In some circumstances, the WRA is likely to require leachate tests to evaluate the available contaminants that can potentially be leached from the sediment.

- Information to demonstrate a benefit to agriculture is likely to be required on a site-specific basis.

- The WRA is likely to require, as a minimum, consultation with the NRA and ADAS or similar who would represent "properly qualified advice" (DoE, 1994; 1995). A specialist report would be preferable and more effective in supporting an exemption application.

- Dredged material disposed of in a confined facility, such as a lagoon, for agricultural benefit is likely to be unacceptable.

5.6   ASSESSING ECOLOGICAL IMPROVEMENT

5.6.1   Introduction

There is no formal definition of ecological improvement within the Waste Management Licensing Regulations, but its legal meaning is discussed in Section 3.5.5. Ecological improvement could potentially offer a route to apply for an exemption on a number of disposal sites but, depending on the WRA's requirement for "properly qualified advice", the scope for exempt disposal activities may in fact be substantially restricted. The legal background to ecological improvement within the Waste Management Licensing Regulations is established as for "benefit to agriculture", and is discussed in Section 5.5.2.

5.6.2   Exemption

The main points to bear in mind when evaluating the disposal of dredged material for ecological improvement are as follows:

- There is no formal definition of ecological improvement. Decision-making on registered exemptions is at the discretion of a WRA.

- There are limits to what may be considered as ecological improvement. Therefore a WRA is likely to require "properly qualified advice" on the nature of the dredgings, the receiving environment and how the application produces an improvement. Properly qualified advice may entail the commissioning of a suitably qualified consultant or specialist.

- The maximum amount of dredged material that can be spread on land (5000 tonnes per hectare per year) may be deemed unsuitable for ecological improvement within the other constraints of the exemptions.

- Each exemption is likely to be assessed by a WRA on a case by case basis.

- The exemption is only effective if it is registered with a WRA by the establishment carrying out the disposal operation and the particulars of the activity are furnished.

Examples of ecological improvement include:

* improving the biodiversity of a site, perhaps by creating different habitats to attract different species

* creating or restoring a habitat which is characteristic of a region's naturalness (e.g. a fen in Fenland)

* creating or restoring a habitat which is scarce or would be regionally important.

5.6.3 Potential uses of dredged material for ecological improvement

The vague definition of ecological improvement presents difficulties in guiding disposal practitioners as to the options open to them under this exemption route. The use of "ecological" in literal terms is potentially much more restrictive than (say) "environmental" would have been. Ecological improvement involves the creation or improvement of plant and animal habitats. However, this phrase is rather imprecise in terms of using dredged material for habitat development or improvement. Habitat development using dredgings is a well established practice in the USA. where material used for such purposes is not even initially viewed as waste. USA equivalents to ecological improvement vary from using dredgings to bury a problem soil, to re-creating islands, intertidal marshes, oyster beds, etc. It is therefore important to discuss disposal proposals with the WRA at the earliest opportunity to ensure both parties' interpretations of ecological improvement are mutually acceptable and are within the terms of the Waste Management Licensing Regulations.

As discussed in Sections 3.5 to 3.7, there are three exempt activities for dredgings disposal as ecological improvement.

* Spreading on agricultural land

There is unlikely to be any ecological improvement from the placement of dredgings on agricultural land unless a significant area (e.g. a field) is to be taken out of production. In such cases any effort to increase the wildlife value of an area may centre on the reduction of the fertility of nutrient-rich agricultural land by allowing an increase in plant biodiversity. The spreading of some dredgings on fertile soils tends to work in the opposite direction, however, either by increasing the amount of nutrients, or by increasing the availability of the existing nutrients to plants, thus promoting the vigorous growth of fewer plants. Overall the scope for this exemption route is small.

* Spreading for land reclamation and improvement

This exemption option offers potentially the greatest scope for the disposal of dredgings under ecological improvement. In the USA dredged material is used to bury contaminated soil, and it is possible that dredgings could be used in the UK to improve or reclaim land in such a manner that ecological improvement is viable. It would be reasonable if the WRA took the view that some form of habitat on a formerly derelict or barren area constituted ecological improvement. It is worth noting that even contaminated or abandoned land can sometimes support valuable plant communities, and so care should be taken to ensure that important species are not affected by disposal despite the existing land condition. In this respect, properly qualified advice is important.

When reclaiming or improving contaminated land with dredged material, reference could be made to the ICRCL threshold trigger concentrations for redevelopment end uses. Concentrations are given for various substances which may be present in dredged material, and an indication of how this could affect the planned use of the land. For example, a planned use such as ecological improvement might mean that the land is classified under ICRCL uses as a "park", "open space" or "landscaped area" (see ICRCL, 1987).

Inherent to ecological improvement is the difficulty in distinguishing it from environmental enhancement or improvement. Reference to English Nature's Natural Area conservation divisions (e.g. see English Nature, 1993a) might help resolve interpretations in terms of guidance on regionally valuable species. Natural Areas encapsulate landscapes of similar ecological value and describe the existing habitat and geology of an area, irrespective of local government boundaries. For example, a Natural Area may include Fenland, Broadland, or Breckland in eastern England. Therefore, an ecological improvement may be demonstrated if vegetation of a Natural Area (or of other regional value) can establish and maintain itself on a disposal site. A Natural Area is not a designated site such as a SSSI.

- Spreading on banks and towpaths

Bankside vegetation is often an important habitat within a local area and therefore potential problems may occur if deposited dredgings smother sensitive plant species. Smothering, particularly with nutrient-rich material, encourages vigorous and competitive plants to establish (e.g. nettles). It is therefore necessary to ensure that no detrimental effects result from the bankside placement of dredgings. Nature conservation organisations may, in fact, discourage the disposal of dredged material to bankside locations.

Prior to the Waste Management Licensing Regulations, many disposal practices involved depositing dredged material on the banks of watercourses. For example, this was normal practice for the Basingstoke Canal Authority, the Broads Authority, and various internal drainage boards. In many circumstances this operation can remain unchanged due to the exemption provisions. However, if the material is to be transported from where the dredging took place (i.e. to other banks) the WRA may require some evidence that such disposal has an ecological improvement to satisfy the qualifications for this exemption.

5.6.4 Statutory consultees

If the disposal of dredged material is to take place within a SSSI, then English Nature, Scottish Natural Heritage, or the Countryside Council for Wales must be approached as a statutory consultee. These conservation organisations should also be involved in consultation for other specified sites such as Special Protection Areas (SPAs), etc. It is unlikely that these organisations would wish to be involved in assessing dredging disposal operations for ecological improvement which do not directly or indirectly affect sites of nature conservation importance, although they may wish to give advice at their discretion. Non-statutory consultees, such as Wildlife Trusts, are likely to be interested and able to give advice on ecological issues and potential improvements on disposal sites outside SSSIs.

Each SSSI citation has a list of potentially damaging operations (PDOs). The disposal of dredgings on a SSSI, even to benefit ecology, would be covered by a PDO. Discussion with the consultee may lead to the requirement for a management agreement. This could be necessary to formalise the disposal of dredgings to a SSSI and satisfy the consultee in terms of ecological improvement.

If the proposed disposal location is situated close to a SSSI boundary there may be some potential for an impact on the SSSI itself. In such circumstances the role of the statutory consultee is less clear. It should be considered preferred practice to approach the conservation organisation, at least to inform them of the proposed disposal activity.

It is likely that the statutory conservation organisations will take a precautionary approach when evaluating any disposal option which may affect a SSSI. For example, consultation with English Nature revealed that they would want to have information passed to them to demonstrate that no adverse impact would result from the spreading of dredged material. This could include the submission of raw data and the analysis and assessment of the disposal operation and its potential problems (e.g. contamination, transportation of material, impact on specific interests, etc).

5.6.5 Summary of requirements for benefits for ecological improvement

A series of investigations and consultations on the ecological improvement of land by means of disposing dredged material has identified the following:

- The definition of ecological improvement can only be site-specific.

- The WRA is likely to require "properly qualified advice" on the beneficial use of dredged material, identifying its ecological improvement. This should include advice on the quantities and frequencies of spreading rates appropriate to the nature of the dredged material, each receiving soil and each receiving site.

- Any uses of dredged material on designated sites must involve consultation with the relevant statutory nature conservation organisation (e.g. English Nature, Scottish Natural Heritage, etc.).

- In some cases, especially for potentially contaminated materials or for disposal on environmentally sensitive sites, some form of environmental appraisal may be necessary as an assurance of no detrimental impacts.

- The WRA should be regularly consulted regarding the spreading of dredged material as an exempt activity.

- It is likely that WRAs would ask for formal reports furnishing the particulars of the exemption, including details of the ecological improvement.

5.7 IMPACTS OF DISPOSING OF CONTAMINATED SEDIMENTS

5.7.1 Introduction

The potential environmental impacts associated with the re-use or disposal of dredged material are a function of:

- the ability of the dredged sediment to cause an effect by virtue of its properties

- the ability of the receiving site to reduce or increase the magnitude of the potential effects, primarily on water quality, by virtue of the site's physico-chemical properties.

The properties of the dredged sediment may be responsible for both short-term and long-term environmental impacts. Short-term impacts are likely to be caused to the surface water quality of nearby watercourses. Such impacts may be a function of the amount of contaminants adsorbed into fine suspended matter in the initial dewatered effluent from the onland disposal or keeping site.

The longer-term impacts of disposal of contaminated dredged material generally relate to the various transport mechanisms by which the contaminants can reach and affect environmental characteristics and enter the food chain. These pathways should, therefore, be examined when a disposal or re-use option is being assessed in order to identify the risk of environmental impacts as a result of the disposal of contaminated dredged sediment. The primary transport mechanisms are:

- surface run-off to adjacent surface watercourses
- leaching into groundwater
- direct uptake by animal population
- plant uptake and subsequent bioaccumulation into the food chain
- direct uptake by humans and animals, for example, of wind-blown fine particulates from a drying disposal site.

5.7.2 Sediment properties affecting disposal to land options

Under the Water Resources Act 1991, it is an offence to pollute controlled waters and the NRA can prosecute those who are found guilty of such an offence. Where disposal to land is considered, some degree of familiarity with the sediment quality and its potential to release contaminants following disposal is, therefore, essential to prevent a pollution incident and identify necessary monitoring requirements. Similarly, information on the sediment quality would be required when applying for a discharge consent to a controlled waterway or a sewer.

When an application for a licensed disposal site is made, the applicant is advised by Waste Management Paper No. 4 to submit a working plan, including information on the quality and quantity of the waste to be deposited. This practice enables other statutory authorities and regulators to ensure that their interests are not affected by the disposal practices. Under the Waste Management Licensing Regulations, the WRA is required to consult the NRA on potential effects on water quality.

When a dredged sediment disposal practice is exempt but there are reasons to believe that the sediment is contaminated, it is likely that the WRA would ask for some sediment quality data, in order to ensure that the disposal option is not likely to cause any effects on the water quality of nearby watercourses or groundwater resources. An operator should therefore contact the relevant WRA(s) to ensure that all their requirements will be included in his disposal application.

The environmental issues of particular concern to the WRA and the NRA will be water and soil quality at the receiving site. Several important sediment properties can influence the impact of dredged material disposal on land. It is these sediment properties and the changes in the contaminants following on land disposal which will determine the potential magnitude of the impacts (Gambrell *et al.*, 1987). These sediment properties include:

- particle size distribution (grain size)
- type and amount of clay
- cation exchange capacity
- organic matter content
- pH and sulphides
- amount of active iron and manganese
- oxidation - reduction conditions.

Given the complexity of the following, qualified advice should be sought in preparation of a disposal application:

- the sediments' properties
- the receiving site's characteristics
- the interactions between dredged sediment and the receiving site
- the chemical reactions which can follow an onland disposal.

In this way, the applicant should be able to demonstrate to the WRA that he/she is aware of the likely impacts following on land disposal and has prepared the necessary mitigating measures based on qualified advice.

5.7.3 Surface runoff to watercourses

Surface runoff from the disposal site can be expected to take place during dewatering in the disposal operation, as well as during rainfall events. Any discharge of contaminated surface water or effluent to controlled waters will require a consent from the NRA. Consent standards for point discharges of effluent from a licensed disposal or keeping site would be set by the NRA to protect the quality of receiving waters. They include maximum allowable concentrations of suspended solids and other substances in the effluent. The standards for contaminants are based on the Environmental Quality Standards (EQSs) for surface waters. NRA may also ask the WRA to set licence conditions relating to the working plan for the site (see Section 2.9.7) whereby runoff is reduced by the method of operation (e.g. progressive restoration and seeding).

It may be possible to direct the runoff from the disposal site into an existing sewer. The discharge consent conditions will be set by the appropriate water company and are likely to be at least as onerous as the NRA standards described above. Conditions for discharge into sewers will vary depending on the capacity of the receiving system and the capability of the authority's treatment works.

The environmental risk of such surface runoff from the disposed or stored dredged sediment depends, in part, on the characteristics of the receiving medium. A possible impact on the receiving water quality and aquatic life from surface runoff is likely to be experienced during the first flush of heavy rain following a long spell of dry weather. In this situation, poor quality runoff can be produced by the leaching of potentially toxic metals from the deposited or stored contaminated dredged sediment. The initial flush may contain significant quantities of metals following the oxidation of sulphides (to sulphates and sulphuric acid). Any disposal system should take this possibility into account.

Similarly, surface runoff with a high suspended solids content can have water quality and ecological implications, particularly if the receiving watercourse sustains a coarse fishery. In such circumstances, it is possible to mitigate the potential impact(s) by constructing land drains or a similar settling facility between the disposal site and the watercourse.

It is possible that metal contamination can occur from natural sources. Dredgings from a watercourse which receives discharge from land drains draining peat soil overlying metal-rich subsoil, or a catchment geology rich in metallic ores and minerals, can have high levels of metal sulphides. Although such an occurrence may be rare, early consultation with the WRA/NRA may prove valuable for selecting the most appropriate disposal option for such dredgings. In addition, the possibility of metal contamination from natural sources compounds the need for a thorough awareness of the characteristics of both the dredged sediment and the receiving site.

5.7.4 Leaching into the groundwater

Under certain geological conditions, for example where there are permeable or fissured strata, the percolation of leachates into deep subsoil layers and even aquifers can occur. Such conditions can lead to the contamination of previously unpolluted groundwater and in certain locations the contamination of water destined for human use and consumption. Care must, therefore, be exercised in selecting the site for the disposal of contaminated dredgings. Groundwater, once contaminated, is extremely difficult and expensive to remediate. Regulation 15 of the Waste Management Licensing Regulations gives effect to provisions of the EC Directive on the protection of groundwater against pollution caused by certain dangerous substances; more details are given in Section 3.12.

The NRA has recently published its policy on the protection of groundwater from sources of contamination. Consultation with the NRA highlighted the Authority's view that the protection of groundwater quality is of primary importance. The NRA has a duty under the Water Resources Act 1991 to monitor and protect the quality of groundwater (Section 84) and to conserve its use for water resources (Section 19). The proposed plan classifies groundwater according to its vulnerability to pollution, and "Source Protection Zones" will be delineated around groundwater sources (e.g. springs and boreholes) for major public water supply sources. Source Protection Zones are determined by the travel time of potential pollutants and the groundwater source catchment areas. Box 5.7.4(a) summarises the Source Protection Zones as defined by the NRA, in their *Policy and Practice for the Protection of Groundwater* document (NRA, 1992).

Box 5.7.4(a)   Source Protection Zones

The risk of pollution to an existing groundwater source is based on the proximity of the groundwater abstraction to the activity likely to give rise to the pollution. The NRA recognise three Source Protection Zones: (a) Zone I (Inner Source Protection) (b) Zone II (Outer Source Protection) (c) Zone III (Source Catchment). In any one situation, the orientation, shape and size of the zones are determined by the hydrogeological characteristics of the underlying strata and the direction of groundwater flow (NRA, 1992).

(a)   Zone I is located immediately adjacent to the groundwater source. The area is defined by a 50-day travel time from any point below the water table to the source and has a minimum of 50 metres radius from the source. This 50-day travel time zone is based on the time it takes for biological contaminants to decay. Consultations with NRA indicated that this is an established standard which is also used in other countries.

(b)   Zone II is larger than Zone I. It is the area defined by a 400-day travel time from any point below the water table to the source. The travel time is based upon the time it takes for the decaying pollutant to reach the groundwater resource while the strata are delaying and attenuating the pollutant.

(c)   Zone III covers the complete catchment area of a groundwater source. All groundwater within it will eventually feed into the source. It is defined as an area needed to support an abstraction from long-term annual groundwater recharge, also termed effective rainfall (NRA, 1992).

The NRA have produced a schematic diagram to illustrate the relationship between the three Source Protection Zones and the groundwater sources. This diagram shows the variation in the relationship between the three zones in four situations where the underlying strata vary substantially. In reality, the size, shape and relationship of the zones will vary significantly depending on the soil, geology, the amount of discharge and the volume of water abstracted. The area of extraction, for example, of a pumping borehole in an aquifer of relatively low effective porosity (e.g. chalk) is greater than in an aquifer with higher storage capacity (e.g. sandstone). This will have the effect that a sandstone aquifer in Zone II is likely to be significantly smaller than Zone III, although chalk aquifer areas will be comparable (NRA, 1992).

Source: NRA (1992)

Box 5.7.4(b) summarises the implications of the nature of overlying soil cover at the disposal site.

The NRA has also engaged upon a programme of mapping groundwater vulnerability in terms of geology and soils at a scale of 1:100,000. These maps will be progressively published from 1994 onwards and will be available from HMSO. The classification of groundwater vulnerability is based upon four key variables:

- nature of overlying soil covers
- presence and nature of drift
- nature of strata
- depth to water table (thickness of the unsaturated zone).

The concept of groundwater vulnerability recognises that the risk of pollution from a given activity will depend on certain hydrological, geological and soil situations. The risk of groundwater pollution from a disposal site receiving contaminated dredged sediment can be assessed based on the nature of the sediment, the natural vulnerability of the groundwater, and the scale of preventive measures proposed to reduce the risk.

Box 5.7.4(b)   Nature of overlying soil cover

Variations in groundwater vulnerability can be recognised on the basis of the physical properties of the soil which affect the downward movement of soil water, and the ability of the soil to attenuate three types of pollutant. These three types are:

- diffuse source pollutants which, under certain circumstances, can be retained in the soil layer (e.g. pesticides); this would be applicable in cases where sediment disposed of on land contains significant quantities of pesticides
- diffuse source pollutants which can readily pass through the soil layer (e.g. nitrate)
- liquid wastes.

The Soil Survey and Land Research Centre (SSLRC) have developed a threefold classification of soil types for the NRA, based on physical soil properties.

- Soils of High Leaching Potential, H: These soils have little ability to attenuate diffuse source pollutants. Non-adsorbed diffuse source pollutants and liquid discharges will percolate rapidly through them
- Soils of Intermediate Leaching Potential, I: These are soils which have a moderate ability to attenuate diffuse source pollutants or in which it is possible that some non-adsorbed diffuse source pollutants and liquid discharges could penetrate the soil layer
- Soils of Low Leaching Potential, L: These are soils in which pollutants are unlikely to penetrate the soil layer because water movement is largely horizontal or these are soils which have a large ability to attenuate diffuse source pollutants. Generally these soils have a high clay content (NRA, 1992).

If soils of low leaching potential are adjacent to soils of intermediate or high leaching potential, it is possible that runoff from the former to the latter may reach the groundwater supply. This fact should be recognised by the applicant in cases where waste disposal licence applications are considered as the WRA is likely to assess an application on the basis of the soil properties in both the proposed receiving and adjacent sites. A full assessment of groundwater resource vulnerability can only be achieved by studying the local soil characteristics and the proximity of pathways likely to facilitate the leaching from a disposal site to groundwater resources.

Source: NRA (1992)

Both the nature of the strata and the depth of the unsaturated zone are associated with the presence of aquifers. However, evaluation of how the nature of overlying soil at a disposal site can influence the pathways of potentially toxic contaminants is difficult.

As the properties of overlying soil vary from site to site, any application for a Waste Management Licence for contaminated dredged sediment will be judged on the disposal site's characteristics. The NRA, in its document *Policy and Practice for the Protection of Groundwater* (NRA, 1992), highlights the need for local studies, including hydrogeological and soil investigations, in order to assess the risk of groundwater pollution from overlying disposal activities. The granting of a Waste Management Licence by a WRA will also be on the basis that no impacts on groundwater quality are expected. Hydrological investigation is deemed necessary by the NRA as part of a waste disposal licence application and the onus is on the applicant to evaluate the impact which the proposed disposal will have and to suggest mitigating measures. An assessment of the investigations and findings will be carried out by WRA officers in conjunction with the NRA as part of the decision-making process before granting a licence. Therefore, undertaking a preliminary study (e.g. a baseline risk assessment) of the disposal site, its soil and underlying strata, and the potential contaminant pathways and targets could prove to be advantageous. In addition, the NRA has indicated that it would encourage operators to contact the NRA departments with responsibility for surface water and groundwater quality at the earliest opportunity in order to discuss their proposals.

The NRA's guidance on the protection of groundwater is primarily aimed at planning authorities, WRAs and other relevant organisations. The NRA is a statutory consultee on disposal site development plans and many aspects of development control under the Waste Management Licensing Regulations. In the case of a licence application to dispose of, store or treat dredged sediment, the Waste Management Licensing Regulations establish an obligation on WRAs to consult the NRA on the potential risk to surface water and groundwater resources, prior to gaining a licence. The NRA, however, could also be approached by applicants for a Waste Management Licence to assist in either identifying or confirming a suitable disposal site.

5.7.5 Direct uptake by animal population

For the majority of contaminants, the risk of livestock being affected by grazing on contaminated soil depends almost entirely on the amount of contaminated soil that is ingested. In general the risk to livestock depends on:

*   levels of contamination in the soil
*   level of soil contamination on the grass sward
*   variations in dietary intake over the year
*   the type, species, age and health of the animals
*   the length of time they are grazing
*   any supplementary food they receive.

The amount of soil contamination on grass pasture will vary with:

*   the type of grass sward
*   its thickness
*   the time of the year
*   weather conditions
*   stocking density
*   grazing management.

It is possible to reduce the risk of direct uptake of contaminants by animals by ensuring that dense pasture is developed on contaminated dredged sediment (MAFF, 1993). For further guidance on risk to livestock grazing on pasture developed on contaminated land, reference should be made to MAFF's *Code of Good Agricultural Practice for the Protection of Soil* (1993). Specialist advice should also be sought if this option is to be considered.

Finally, care should also be taken in using dredged material as long- and short-term effects on livestock can occur as a result of any debris (e.g. sharp objects) in the sediment when spread on agricultural land.

5.7.6 Plant uptake and subsequent bioaccumulation in the food chain

As well as direct ingestion of contaminated soil, livestock could be affected by toxic contaminants which have been taken up and accumulated by the vegetation on which they graze. This is an indirect path of poisoning and is, therefore, regarded as less significant than ingestion. Nonetheless, there are several metals (e.g. cadmium, copper, boron, nickel and zinc) which are toxic to plants (phytotoxic) at significant concentrations. Some potential contaminants are of particular relevance to crop plants and the food chain (e.g. cadmium and lead). Again, properly qualified advice should be taken on the susceptibility of different plants to the uptake of contaminants.

5.7.7 Wind-blown dust from disposal sites

As the dredged sediment dries out at the disposal site, wind erosion of the dried material can facilitate the airborne transport of adsorbed contaminants. Also, some gaseous or volatile emissions may be expected to occur during and after the disposal operations. Both these pathways can potentially be responsible for a direct uptake of, or contact with, potentially harmful contaminants by animals and humans. This possibility is of particular significance when such sites are in the proximity of residential and recreational areas, or if they are near intensively used public footpaths or towpaths.

Prior to identifying a disposal site, the uses of both that site and neighbouring areas should be identified. For instance, the need to redirect public footpaths and the distance from the disposal site should be clarified in advance. The Highways and Recreation Department of the relevant Council should be notified, and the disposal plans and associated footpath redirection proposals and/or temporary diversion orders confirmed with them (see Section 3.7.3). An operator should contact WRA about covering contaminated disposal sites, as WRAs may require covering with uncontaminated material allowed by progressive restoration of the site.

Encouraging the establishment of vegetation following disposal could help to prevent wind-blown dust, but in cases it may be necessary to assist the growth of vegetation because dredged material properties may otherwise inhibit or retard plant growth. Restoring a disposal site has the added advantage that surface vegetation minimises ingress of water and subsequent leaching.

5.8 IMPACTS OF DISPOSAL OF UNCONTAMINATED SEDIMENTS

In addition to the impacts of contaminated dredged sediment disposal discussed above, the disposal of uncontaminated sediments can also have environmental impacts. When spread on agricultural land, for example, uncontaminated sediment can have negative impacts on the soil structure and on nearby watercourses if it is saline or if it contains high levels of fines. Disposing of dredged material on a flood plain also requires careful consideration.

5.8.1 Saline dredged sediment

Even though salinity is not a direct contaminant, sodium ions can interfere with the structure of agricultural soils. When this happens, soils become very difficult to cultivate. Soils with high silt or clay contents are particularly vulnerable in this respect. Peaty and coarse sandy soils, on the other hand, are less prone to deterioration from sodium ions.

In fine sediments, it is possible to mitigate such effects by increasing the speed of drainage from the disposal site. In freely-draining soils, rainfall should contribute to the rapid removal of sodium ions from the dredged sediment. In some cases dry underlying soil is likely to speed up the downward movement of saline water.

The effect of saline water on the soil structure can similarly be minimised if dredged sediment is mixed with estuarine/river water prior to pumping ashore, diluting the sodium ion content.

Runoff water from saline dredged sediments following onland disposal is likely to contain sodium ions. The introduction of these ions into surface water or groundwater may interfere with the function of these water resources. The advice of the NRA should therefore be sought in advance.

When identifying disposal sites for saline dredged sediment the following characteristics should therefore be considered in the pre-disposal working plan:

- content of sodium and chloride ions in the dredged sediment
- the type of soil (e.g. peat or clay/silt) of the receiving agricultural field
- the possibility of diluting the sodium and chloride content prior to disposal
- the possibility of speeding up drainage following disposal
- the proximity of surface water and groundwater resources and their functions (see, for example, Section 5.7.4); early consultation with WRA/NRA should identify any such resources
- the proximity of important fresh water habitats with little tolerance to salinity fluctuations.

With regard to the proximity of surface water and groundwater resources and functions, NRA will advise on whether any of their Source Protection Zones (Section 5.7.4) may be affected, whether or not disposal should therefore take place, and under what conditions. If disposal does take place in such areas, frequent monitoring of the watercourses will almost certainly be required.

NRA and English Nature should be contacted to identify whether or not any fresh water habitats vulnerable to salinity fluctuations are at risk and what mitigating measures should be applied.

5.8.2 Dredged sediment containing high fines content

The application of dredged sediment with a high fines content may be beneficial to agricultural fields as a soil conditioner. The runoff water, in particular the initial dewatering effluent, however, may contain high levels of suspended solids.

The runoff of this effluent into surface watercourses could have a significant impact if the watercourses sustain aquatic life susceptible to high suspended solids, or if they are used as a drinking water source. For example, the NRA would be concerned about the potential for spreading of dredged sediment with high content of fines to affect a coarse fisheries habitat.

Under the Waste Management Licensing Regulations, when dredged sediment is spread on sites for the purpose of either agricultural benefit or ecological improvement, an exemption must be registered, but there is no obligation on the WRA to consult with the NRA. The NRA do not have any statutory powers to prevent the deterioration of water quality and interference with the uses of the water resources, but it can prosecute those who cause such deterioration. It is therefore advisable to contact the NRA in advance of any spreading of dredged sediment near to a river, stream, land, lake or reservoir, even if that sediment is not contaminated.

5.8.3 Impacts of disposing on flood plains

Where dredgings are to be spread on land in the floodplain, NRA will wish to consider what impact the dredgings will have on the available flood storage and conveyance. This will be particularly important in areas with clearly identified washlands and/or where there is currently a low standard of protection against flooding to property. The local planning authority will formally seek NRA's views where planning permission is sought for such sites (such as in connection with the exemption of reclamation of land; see Section 3.6). Any work, including storage and disposal, within the flood banks of a designated main watercourse will require NRA consent under bye-laws. Such consent will be refused if the proposed works are detrimental to flood defence interests (i.e. if they have significant adverse affect on the channel and/or floodplain capacity).

Again, it is advisable to consult the NRA (Flood Defence) prior to disposal. In particular, it would benefit the operator and aid the consultation with the NRA if the NRA were made aware of proposals with GDO exemptions early in the decision-making process. Failure to consult has resulted in problems in the past.

5.8.4 Habitat creation

Finally, as indicated earlier, the disposal of dredged material, even at a licensed site, can have significant medium- to long-term benefits for nature conservation by creating new habitats (e.g. lagoons) which are largely undisturbed by human activity. While this sort of impact is largely immaterial to the Waste Management Licensing Regulations, it is nevertheless important in providing a balanced overview of the environmental effects of disposal.

6 Re-use, storage, keeping and disposal of dredged material

6.1 INTRODUCTION

This chapter examines and gives examples of the options which may be available not only for the re-use or disposal of dredgings but also for the storage and keeping of the material. The optimum combination of storage and re-use or disposal options will need to be found for each dredging operation. The use of selective dredging where feasible may result in significant cost savings in any re-use or disposal option (for example in minimising the quantity of contaminated material to be paid for when disposing at a licensed tip).

An important consideration for many operators will be the differences between current storage, disposal or use practices and those which are deemed acceptable under the Waste Management Licensing Regulations. Current practice will usually fall into one of the four new categories:

- not a waste
- waste but exempt from licence requirements
- waste acceptable if licensed
- waste not acceptable even if licensed.

Discussions with the WRAs and the NRA have suggested that the existing conditions at (or the characteristics of) the receiving site for use or disposal will be an important consideration in determining (or proving) the acceptability of a particular option.

The sections which follow describe the various options available and, for exemptions in particular, attempt to answer questions which may arise in relation to current practice.

6.2 OPTIONS FOR STORAGE

6.2.1 Why store?

Storage of dredged material prior to its final disposal or use is a common practice since direct disposal is not always feasible. Depending on the intended end use, the period for storage will vary. Dredged material may be stored for several reasons, for instance:

- the material is waiting to be spread on land in connection with the reclamation or improvement of that land but the development of the receiving site is insufficiently advanced, or the conditions required for spreading (on agricultural land, for example) are not yet favourable

- the dredged material is to be sold but the buyer is not able to use it yet or a buyer has not been identified

- the material needs to be bulked before loading into larger vehicles for onward transportation

- a location has been found for the disposal or use of the dredged material but, before transport takes place, drying or other treatment of the material is essential to reduce the costs associated with transport or disposal

- a treatment plant for dredged material is in operation (e.g. for selective separation into different grain sizes) but the material to be treated arrives at a much faster rate than the treatment process, subsequently creating a stockpile of material to be treated.

Where the dredgings are waste (see Section 3.2) a storage site will require a Waste Management Licence unless it is part of an exempt activity (see Section 3.3) and/or is undertaken in the following circumstances:

- storage at the location where spreading on agricultural land will take place (see Section 3.5.6)

- treatment involving screening or dewatering at the bank or towpath where the dredging operation takes place, or at the place where the dredgings are to be spread on agricultural land (see Section 3.5.7), in connection with land reclamation (see Section 3.6), and on banks and towpaths (see Section 3.7).

A number of factors such as technical feasibility, practicality, legislation, environmental constraints, and, in particular, economics, will determine whether a material can be stored. Dredged material may be stored on land or, less commonly, under water.

6.2.2 Storage on land

When dredged material is stored on land, the material will be in a drying environment and leaching of possible contaminants from the material will start almost immediately. In many cases the drying out of the material will, in fact, be one of the objectives. Knowledge of the characteristics of the material (see Section 4.1) should therefore be used to determine the extent of protection measures needed to ensure that any contaminants will not cause pollution.

Table 6.2.2 sets out criteria on technical feasibility, together with environmental considerations and legal constraints, and demonstrates some of the advantages and disadvantages of storing dredged material on land.

6.2.3 Waste transfer stations

Waste transfer stations are sites usually remote from the dredging operation where waste is temporarily stored pending transportation to a permanent disposal site. A licence will be required for such a site under the Waste Management Licensing Regulations covering aspects such as maximum quantity to be held and monitoring requirements. The characteristics of such sites are the same as those for storage on land as summarised in Table 6.2.2.

6.2.4 Storage under water

Storage of dredgings may sometimes be possible in gravel pits or at other locations in waterways where navigation will not be disrupted by the storage of dredged material. Storage in gravel pits and the like will require planning permission if there is a change of land use, and a Waste Management Licence if there is no beneficial use determined for the material. In addition, a statutory consent may need to be obtained by application to the responsible authority for the watercourse, when the channel profile is to be temporarily altered. In assessing the possibilities of storing dredged material in this manner, the quality of the material will be an important determinant of whether the option can be justified. As discussed in Section 5.7, the NRA has statutory powers to prevent actions which may have a polluting effect on the water environment. If the dredged material is "inert" and an assessment has been made to ensure that it will not have any negative impacts on its direct environment, this form of storage may be an economically attractive option. However, the suspension of fine particles within the water column may adversely affect other water users, and care must therefore be exercised when undertaking such operations.

Table 6.2.4 sets out examples of the criteria determining technical and economic feasibility, and highlights some of the environmental considerations and legislative constraints together with the advantages and disadvantages associated with this form of disposal.

Table 6.2.2   Storage on land

Option	Technical feasibility	Environmental considerations	Legislative constraints	Advantages	Disadvantages
Storage on land	- Sufficient space should be available for the quantities of dredged material to be stored. - Some form of dewatering mechanism should be provided to encourage the drying of the material. The installation of a temporary drainage system beneath the storage area; a large surface with a minimum thickness of the stored material; and frequent rotovating of the material are all measures which will help to increase the speed of the drying process. - Suitable material is required for any bunding. Depending on volume of stored dredgings, bunding may require a temporary works design by a qualified geotechnical engineer.	- A site assessment should be carried out to establish any impacts on the local geology and hydrogeology if the storage time is long. - Dust and noise should be kept to a minimum when material is handled. - Any public rights of way should be protected and maintained, or arrangements made for formal diversions. - Contaminated or silty water must not be allowed to drain back to the watercourse.	- No Waste Management Licence is required for storage operations if: • the material is stored adjacent to where the dredging was carried out • the material is to be disposed onto agricultural land for beneficial use and/or ecological improvement • the material is to be used beneficially, without any further treatment. - A Waste Management Licence will be required by the WRA if the storage of the material does not comply with the exemptions or is not to be used beneficially. - Will require planning permission if a Waste Management Licence is required or there is a change of land use. - A (temporary) discharge consent may be needed from the NRA.	- In the case of selling the dredged material, storage of the material gives an opportunity to wait until an appropriate buyer is found or an attractive selling price is negotiated. - Drying of the material will continue during the storage process, enabling the material to be handled and transported by standard earthmoving equipment more cost-effectively. - Drying of the dredgings will often reduce the volume of the material and hence reduce transportation costs and the amount of final disposal area required.	- Site preparations and possible protective measures needed to ensure that no contaminants leach in to the groundwater or an adjacent watercourse may be expensive. - Protective measures for public safety, such as fencing, need to be checked and maintained regularly throughout the storage period.

Table 6.2.4 Storage under water

Option	Technical feasibility	Environmental considerations	Legislative constraints	Advantages	Disadvantages
Storage under water	- Access needs to be available to facilitate the subsequent removal of the material. - Thought should be given to the required minimum depth of water over the stored material in order to determine the machines for later dredging. - The storage area needs to be free of significant currents to ensure the material remains in place.	- An assessment will need to be made on the impacts of dredged material storage on the water column since certain chemicals may become available after disturbance, resulting in a deterioration of water quality due to contamination. - If the location is used for navigation or recreation a minimum depth after disposal will need to be determined. - An assessment will be required to ensure no existing bed life is damaged or eliminated by storage.	- Waste Management Licence will be required. - Planning permission might be required if there is a change of land use.	- Placement of the dredged material to be stored can be fast and cheap. A hopper barge can be used if access via water is available; in other situations pumping might be possible. - No visual impact is created on land since the material remains under water.	- The material will not undergo a drying process, therefore subsequent drying may be necessary prior to disposal or beneficial use. - There will be a loss of volume of the water body in which the material is placed. - Double handling of the dredged material cannot be avoided. - There is a risk of water pollution occurring.

6.3 RE-USE OPTIONS

6.3.1 Introduction

The following options for re-use have been identified as a result of consultation and a review of available literature. The suitability of each option is obviously dependent on local markets and any financial advantages or disadvantages.

- Use in landscape or reclamation projects including golf courses, landfill sites, etc. as covering material. Sandy/loamy material is specifically favoured for golf courses when it is incorporated in to the top layer, as it improves the bearing capacity of this layer during wet periods. Much larger materials like gravels can be used as drainage fill.

- Road construction; dredged material can be used only if the material is of an appropriate particle size.

- Use in industry; for example in the manufacture of bricks, ceramics, tiles or artificial gravel, or as low-quality construction material. In most cases the material needs to be clean and of an appropriate particle size, without any larger debris.

The technical, environmental and legal considerations together with the advantages and disadvantages of re-using dredged material as outlined above are given in Table 6.3.1.

Other re-use options which may be available are:

- beach nourishment
- river enhancement.

6.3.2 Beach nourishment

Given suitable environmental conditions and the correct type of dredged material, beach nourishment could potentially utilise large quantities of dredged sediment. Beach nourishment involves supplementing the natural supply of beach material by artificial means. Beaches, for the purpose of this document, include not only sand and shingle foreshores but also those consisting of silt, the latter being common in more sheltered areas such as estuaries.

Beach nourishment plays an important role in coastal defence and also assists in improving the environment, particularly for improving the amenity value of existing beaches or offering opportunities (particularly in establishing saltings) for habitat creation. Beach nourishment on the UK coast, as defined by the Coast Protection Act (1949), requires approval by the Coast Protection Authority, which is normally the maritime district council. The disposal of dredged material below the level of mean high water spring tide for beach nourishment purposes also requires a licence from MAFF under the Food and Environmental Protection Act (1985). The NRA should also be consulted, as well as other relevant agencies such as local port authorities, English Nature, the Countryside Commission and the Crown Estate (who are generally responsible for the "sea bed" below low water).

Beaches are naturally mobile. A fundamental requirement of successful beach nourishment schemes is a good understanding of the sediment transport regime. This often, but not necessarily, leads to a requirement that the beach nourishment material should consist of sediment which is of a size similar to, or slightly larger than, that which naturally occurs on the beach.

Table 6.3.2 lists the technical and environmental considerations, together with some of the advantages and disadvantages of using dredged material for beach nourishment.

Table 6.3.1   Re-use of dredged material

Option	Technical feasibility	Environmental considerations	Legislative constraints	Advantages	Disadvantages
Re-use	- A thorough assessment of the material is vital, together with undertaking (market) research in order to determine the technical and economic viability of its future (beneficial) use. - Separation of dredgings may be necessary, involving screening at the dredger and separate transportation. - The possible treatment methods necessary before a material can be used need to be considered. This may require the washing of dredged material (for example before it can be used in the ceramics industry). - Storage facilities need to be considered. Quantities arriving may be too large to handle for a particular intended use/process.	- When the material is used for road construction or landscape purposes, the quality of the material is paramount. If there is any doubt about the quality of the material (e.g. slightly contaminated with heavy metals) the possible effects on the surrounding environment need to be assessed. If the material is sold, the responsibility for record keeping transfers to the buyer, although transfer notes should be kept by all parties for two years. - When treatment techniques are used, either to wash or separate materials, consideration needs to be given to the effect of the possible waste products which might arise from such a process.	- If treatment of the material is necessary before it can be sold, the treatment practice may require a Waste Management Licence.	- The economic benefits gained from selling the material may partially or fully cover the costs of dredging. - Volume reduction may reduce the costs of transportation or disposal.	- Storage facilities may be necessary if treatment is required prior to re-use. Associated costs can therefore make the practice expensive. - Selective dredging may be necessary to obtain the right materials for the intended use, therefore making the dredging method more expensive.

Table 6.3.2   Beach nourishment

Technical considerations	Environmental considerations	Legislative constraints	Advantages	Disadvantages
- Sediment size will influence its mobility and thus the extent to which it is moved by waves and currents. - Sediment should be clean to avoid polluting coastal waters. - Sediment should normally be of a similar composition to native material and should be durable. - Wave and current climate should be assessed in detail to define how the sediment will be transported, at what rate it will be lost from the site, and where it will be transported. - Length of time nourishment will last (how fast the material will be lost) needs to be assessed. - Whether further dredgings will be available to re-nourish needs to be assessed.	- Sediment size, nature and quality should be considered in relation to possible impacts on local flora and fauna. - Sediment size, nature and quality should be considered in relation to any existing beach use (e.g. strolling, boat launching, tourism, fishing). - The implementation/construction phase should be considered carefully in terms of interference with beach access, foreshore and nearshore use (fishing and bait digging) and increased turbidity (impact on flora, fauna, fishing). - Possible impacts elsewhere as a result of sediment losses should be defined and the need for control structures (e.g. groynes) identified. Impacts might include nuisance (wind-blown materials), deposition on the sea bed elsewhere, damaging flora and fauna, or deposition in navigation channels. - Future monitoring of coastal process and the effect on the environment should be considered.	- Requires FEPA Licence from MAFF. - Requires approval from Coast Protection Authority and consultation with other relevant agencies such as the NRA, English Nature, etc.	- Can satisfy an existing need in terms of coastal defence and environmental enhancement. - Can potentially be a disposal option which is very simple to execute. - Can sometimes accommodate very large quantities.	- Requires thorough study of coastal processes to assess possible impacts. - Requires tight control on material type (size, nature, quality). - Requires easy water access to site.

6.3.3 River enhancement opportunities

As with beaches, opportunities may exist on inland waterways to use dredged sediment to improve or maintain existing flood defences. Early consultation with the NRA, particularly in areas where earth embankments are located in the flood plain of rivers, may reveal the need for suitable material (clay, sand, silt) to fill low spots or thicken or raise the defence as required.

Other opportunities may exist to use dredged gravel and rocks to fill scour holes in the river bed, particularly close to river control structures.

Environmental uses may include the relocation of gravel arisings to elsewhere on a river to provide spawning grounds for fish. Environmental enhancement may similarly result from habitat creation schemes in low-lying areas adjacent to a waterway. This may constitute ecological improvement but "properly qualified advice" should be taken. As with beach nourishment, English Nature and the Countryside Commission should be consulted when planning to use dredged material in this manner.

Table 6.3.3 lists the technical and environmental considerations, together with some of the advantages and disadvantages of using dredged material for river embankment.

6.4   TREATMENT TECHNIQUES

Before any dredging commences, the material should be assessed for its chemical composition, physical properties and moisture content (see Chapters 4 and 5). On the basis of this assessment, the potential suitability of the material for re-use can be established. If the material cannot be used in its present state, treatment techniques may need to be applied. Possible treatment techniques, as discussed in Csiti (1993), are shown in Table 6.4.1.

Table 6.4.1   Treatment techniques

Treatment technique	Objective	Comments
Natural dewatering (e.g. on river bank or adjacent field).	Volume reduction	The method is cheap and simple but takes space and time.
Mechanical dewatering (e.g. sieve belts, chamber filters, drum sieves and centrifuges).	Volume reduction	Many techniques are available but cost-effectiveness needs to be considered.
Gravity beds	Volume reduction; separation of fine and coarse fraction.	Can be used in combination with dewatering beds.
Hydrocyclones (separation of coarse sand and fines/organic material).	Volume reduction; separation of contaminated fraction.	Only suitable for separation of certain grain sizes.
Extraction (e.g. using acids or flocculating agents).	Removal of metal contaminants.	High removal efficiencies. Selective and non-selective extraction possible.
Concentration techniques; extraction and washing.	Removal of organic contaminants.	At present used for highly contaminated materials, in relatively small quantities.
Destruction techniques (i.e. biodegradation and incineration).	Removal of organic contaminants.	As for concentration techniques.

Note: Information to produce this table has been taken from Csiti (1993)

Once treated, the material will either be disposed of or re-used in a manner which is likely to have been pre-determined through technical, economic and environmental analyses.

Table 6.3.3   River enhancements

Option	Technical considerations	Environmental considerations	Advantages	Disadvantages
River Enhancements	- Characteristics of the material to be dredged will need to be ascertained to enable geotechnical evaluation of potential use for strengthening or topping-up flood defences. - Calculations and plans will be required for approval and possible land drainage consent of the NRA. - Pre-and post-surveys of any scour holes to be filled will probably be required. - Information on particle size and the grading of gravels and sands will be required to justify use for fish spawning grounds.	- Consultation required with interested nature conservation bodies, including English Nature. - Arrangements to be made to re-direct any public footpaths through disposal areas whilst work takes place.	- A use for the material may be found close to the dredging site, reducing transportation and eliminating disposal costs. - Payment for suitable material may off-set costs of dredging.	- Dredged material may require separating with a part going to disposal. - Lengthy consultation may be required with the NRA and other bodies to identify suitable sites.

Although the characteristics of the receiving environment and the dredged material itself will play a role in the decision-making process, it is likely that the financial and economic implications will also be of paramount importance (see Chapter 8).

6.5 INTRODUCTION TO DISPOSAL OPTIONS

Disposal sites are not only used for contaminated material: they are often required where no alternative beneficial use can be found for the dredged material. The following disposal options have been identified as those commonly used by practitioners:

- disposal to agricultural land
- land improvement (e.g. spreading)
- bank disposal
- composting for subsequent spreading
- lagoons
- landfill sites.

Another option which may be available for disposal of dredgings is sea disposal.

As explained in Chapter 3, within the Waste Management Licensing Regulations a distinction can be made between the disposal options known as exemptions and disposal under licence. Sections 6.6 and 6.7 set out the key factors associated with these various disposal options.

6.6 EXEMPTIONS

6.6.1 Introduction

Under the Waste Management Licensing Regulations, several paragraphs deal with the disposal of dredged material through an exempt activity. The applicable paragraphs are discussed in Chapter 3 of this guidance document. For the operators or managers of inland or tidal waterways there is a strong incentive to assess the possibility for exempt activities (if no beneficial use can be found for the material to be dredged). For example, although an exemption must be registered with a WRA, no Waste Management Licence will be required and costs (as compared to the costs of disposal to a licensed site) will be minimised. Conversely, whereas in the past many activities were automatically exempt, under the new Regulations there is a requirement on the operator to notify the registration of each exemption (see Section 3.4).

Box 6.6.1 presents examples of the information requirements of three WRAs for notification of exemptions. It is apparent that the information requested may exceed the requirements of the Regulations, for example, Example C, 4(b); although the agreement of the landowner/occupier must be obtained before undertaking the works, details of the landowner/occupier are not required to register an exemption. Similarly, the Regulations do not require quantities to be given per week for all exemptions (Example C, 6). Operators notifying a WRA of the registration of an exemption should ensure that the quantities given are in line with the requirements of the specific exemption in question. Example A is in keeping with the requirements of the Regulations.

The statutory duty on a WRA is to register any exempt activity notified to it, rather than to "grant" an exemption. Therefore, if an organisation claims an exemption to which it is not entitled it runs the risk of prosecution for the unlicensed disposal of waste (see Section 3.4). To this end, if any one of the factors pertaining to an exemption (see Sections 3.5 to 3.8), including the relevant objectives of the EC Directive (Waste Management Licensing Regulations Schedule 4 Part 1 paragraph 4(1)(a), see Section 3.3), are not met, the exemption may not apply.

Table 6.6.1 lists the technical feasibility, environmental considerations and legal constraints together with the advantages and disadvantages of the option of obtaining an exemption.

Box 6.6.1 Examples of WRA information requirements for notification of exemption

Example A:	Adapted from Humberside County Council WRA (1994)

REGISTRATION OF ACTIVITIES EXEMPT FROM WASTE MANAGEMENT LICENSING UNDER THE WASTE MANAGEMENT LICENSING REGULATIONS 1994

Name and address of the establishment or undertaking proposing to carry out the exempt activity.

Describe the activity which constitutes the exempt activity*. Include any plans and details which may assist the WRA in establishing the basis of the claimed exemption.

Paragraph Reference No:

Place where the activity is carried out:

* This may be the schedule and paragraph references from the Waste Management Licensing Regulations 1994.

Waste management activities exempt from Licensing must be carried out without:

(i)   risk to water, air, soil, plants or animals; or

(ii)   causing nuisance through noise or odours; or

(iii) adversely affecting the countryside or places of special interest.

Example B:	Adapted from Nottinghamshire County Council WRA (1994)

Waste Regulation Authority	Reg No. Date Received Officer

THE ENVIRONMENTAL PROTECTION ACT 1990
WASTE MANAGEMENT LICENSING REGULATIONS 1994
REGISTRATION OF EXEMPT ACTIVITY

The above Regulations exempt persons carrying out various waste related operations from the requirement to hold a Waste Management Licence under the Environmental Protection Act 1990 (see Schedule 3 for details). However, an establishment or undertaking carrying out activities involving the recovery or disposal of waste shall be entitled to rely on an exemption only if registered with the WRA for the area in which the activity takes place.

This form **should not be used** for notification regarding activities prescribed in paragraphs 7(3)(c) of Schedule 3 of the above Regulations (land spreading of waste). **Complete all sections.**

1. ESTABLISHMENT UNDERTAKING THE EXEMPT ACTIVITY

    Name (organisation, company, partnership, authority, society, club, charity etc.):

    Address:

    Post Code:

    Telephone Number: Fax Number:

2. ADDRESS OF FACILITY WHERE THE ACTIVITY IS INTENDED TO BE UNDERTAKEN

    Name

    Address:

    Post Code:

    Telephone Number: Fax Number:

3. ACTIVITY TO BE REGISTERED AS EXEMPT (please quote the relevant paragraph from Schedule 3 of the Regulations)

    ACTIVITY   WASTE TYPE   SCHEDULE 3 PARAGRAPH

4. DECLARATION

I certify that the information in this application is correct and I hereby notify you of the registration of an exemption from the requirement to hold a Waste Management Licence under the Environmental Protection Act 1990.

Name (block letters)   ....

Signed   ....   Date   ....

On behalf of   ....

The provision of false information to register an activity as exempt from the requirement to hold a Waste Management Licence may mean that the operations being carried out do require licensing. Carrying out licensable activities without a licence is an offence under the Environmental Protection Act 1990. The penalty for such offences on summary conviction may be imprisonment for up to 6 months, a fine not exceeding £20,000 or both. On indictment the penalties are up to 2 years imprisonment, a fine or both.

This form should be returned to:

Box 6.6.1 Examples of WRA information requirements for notification of exemption, continued

Example C:	Adapted from Leicestershire County Council WRA (1994)

Register No.
WASTE REGULATION AUTHORITY

Environmental Protection Act 1990
Waste Management Licensing Regulations 1994
Notice of Intention to Rely on Exemption

The above Regulations exempt persons carrying out various waste related activities from the requirement to hold a Waste Management Licence under the Environmental Protection Act 1990. However, an establishment or undertaking carrying out specific activities shall be entitled to rely on an exemption only if registered with the WRA for the area in which the activity takes place. The Regulations require registration to be effected by the establishment or undertaking in question notifying the WRA in writing of its intention to rely on an exemption.

1   Name of establishment:

    Address:

    Post Code:

    Telephone Number: Fax Number:

2   Establishments' trading names:

3   Address of facility where exempt activity is intended to be undertaken:

    Post Code:

    Telephone number: Fax number

4(a) Are you the owner/occupier of the land where the activity is intended to be undertaken?   YES/NO

4(b) If you are <u>not</u> the landowner/occupier, give the following details.

    Landowner's/occupier's name:

    Address:

    Post Code:

    Telephone number: Fax number:

    N.B. You must attach written confirmation of landowner's/occupier's consent for proposed activity.

5   Specify the activity and types and quantities of waste material to be delivered to the site in tonnes per week:

    ACTIVITY   WASTE DESCRIPTION

6   For all the waste types specified in Question 5 above, state the total quantity likely to be delivered per week*
    and kept at any one time:

WASTE DESCRIPTION	AMOUNT DELIVERED PER WEEK*	MAXIMUM ON SITE AT ANY TIME

    PLEASE NOTE: all quantities should be related to the requirements of the specific exemption

7   I certify that the information in this notification is correct and I hereby register an exemption from the

    requirement to hold a Waste Management Licence under the Environmental Protection Act 1990.

    Signed   ....   Name (please print)   ....

    On behalf of   ....   Date   ....

Table 6.6.1 Exemptions

Option	Technical feasibility	Environmental considerations	Legislative constraints	Advantages	Disadvantages

Exemption	- Appropriate machinery needs to be chosen if dredging is carried out from the bankside.
	- When machinery is used for spreading onto agricultural land, care should be taken not to damage the existing soil profile, therefore, careful choice of plant is paramount.
	- On sloping agricultural land it is necessary to prevent water and silt flowing back to the watercourse and temporary storage of part of the dredgings may be required to allow drier material to eventually spread immediately adjacent to the watercourse.
	- When bank side disposal takes place, quantities need to be taken into consideration so that the bank will remain stable and will not undergo any significant changes in strength.
		The Code of Good Agricultural Practice for the Protection of Soil provides general guidance on practices which will maintain or improve the ability of soil to support plant growth. The Code of Practice is not, however, a statutory code.
		- With respect to the potential use of dredged material to improve agricultural land, the following items are of significance:
		  • compaction risk
		  • background information on the receiving agricultural field
		  • possible contamination sources to which the dredged material may be exposed.
		- When "waste" is applied to agricultural land the potential effects on groundwater and surface water need to be considered. MAFF has written a Code of Good Agricultural Practice for the protection of Water, supported by the NRA. This document can be used as guidance to assess what the WRA and particularly the NRA are likely to require.
		- Silt-laden water must not be allowed to drain into the watercourse.
			- Limits on quantities are set defining maximum amounts of dredged material which can be disposed of under the various exemptions.
			- Although no licence is required, the exemption has to be registered and justification is likely to be requested by the WRA regarding beneficial use or ecological improvement.
			- If the specified quantities are exceeded a Waste Management Licence will be required otherwise there is a risk of prosecution by the WRA for the unlicensed disposal of waste.
				- The material does not fall under the Fees and Charges Scheme and therefore licence fees do not have to be paid for the site.
				- If the material can be disposed of close to the dredging site, transport costs will be reduced.
				- Disposal can often be carried out from the waterfront directly onto the receiving area, making road transport unnecessary.
					- It may be a lengthy process to convince the WRA that the material is exempt.
					- Sampling costs and possibly consultants fees may be incurred.

The following text describes the exemptions, supported by questions and answers, in an attempt to give a clear understanding of which practices are exempt under the new Regulations, together with the views of (and examples provided by) WRAs, the NRA and practioners during consultation and site visits.

6.6.2 Agricultural land

This practice involves spreading onto agricultural land for benefit to agriculture or ecological improvement. It is limited to an amount of 5000 tonnes of waste per hectare in any period of 12 months. The operator needs to confirm that there has been no earlier spreading by someone else or there is a risk of inadvertently exceeding annual tonnage (see Section 3.5).

Question	Answer
Dredgings have always been disposed to farmland after agreement with the farmer. Is this going to change?	Not necessarily. But the intended work will need to be registered at the appropriate WRA before being carried out.

Question	Answer
What kind of proof will the WRA require?	It is likely that, particularly in areas where there is doubt about the dredged material, the WRA will require: • compositional analysis of the dredged material (see Section 5.5) • background information concerning the field where the material is going to be spread, such as existing and future usage, nearby watercourses, whether the site has been used for other disposal, etc. • a report from a competent person(s) who can give qualified advice • the quantities involved.

Question	Answer
Providing the proof appears very complicated and very expensive, is it?	It is a little complicated but it is also important. It is very easy to unintentionally cause environmental degradation, for example by polluting groundwater or surface watercourses. To prevent possible pollution, some information is necessary on the quality of the dredged material (see Section 4.4). Obviously some expenditure will be incurred, but with a logical sampling strategy this can be kept to a minimum (see Section 4.3).

Question	Answer
What happens if an exempt disposal is not registered?	Disposal without registration, in the view of the WRA, would represent an illegal practice. Failure to register may result in a criminal prosecution with a maximum fine if found guilty of £10.

Question	Answer
What is the best approach to register an exemption?	Consultation at an early stage with the WRA, before the dredging operation takes place, is the best way forward. Every case will be looked at from a site specific point of view. If given sufficient time in advance, the WRA may ask for information from the NRA or have vital information themselves. This will help them determine whether further information is needed to support exemption. If the WRA are unable to respond quickly, direct contact with NRA may help identify any key issues on problems with the proposed exemption.

Box 6.6.2 provides two examples of the possible information requirements of WRAs for notification of the intention to spread waste on agricultural land. Example A reflects the basic requirements of the Regulations, while Example B has been developed to consider additional information outside the requirements of the Regulations (e.g. 6, 7 and 8). NAWRO intend to develop Example B into a national format.

Box 6.6.2 Examples of WRA information requirements for notification of spreading to agricultural land

Example A:	Adapted from Humberside County Council WRA (1994)

THE WASTE MANAGEMENTLICENSING REGULATIONS 1994
REGULATIONS 1(3), 17 & 18
PRE-NOTIFICATION & REGISTRATION OF EXEMPTION
SPREADING OF WASTE ON LAND USED FOR AGRICULTURE

Name and address of the establishment or undertaking carrying out the spreading:

Telephone Number: Fax Number:

Description of the waste:

Process from which waste arises:

Estimate of the quantity of waste to be spread (single spreading):

OR

Estimate of the total quantity of waste to be spread during the next six months (regular or frequent spreading):

Where the waste is being or will be stored pending spreading (if applicable):

Location of spreading:

Intended date of spreading (single spreading):

OR

Frequency of spreading (regular or frequent spreading):

NOTES:

If the distribution of waste over the land is such that the volumetric loading is substantialy greater on some areas than others greater details should be provided.

The location given should be of sufficient detail to easily identify the site. Alternatively, the area of deposit may be accurately marked on an Ordnance Survey 1:2500 metric edition plan and submitted with this return.

An establishment or undertaking carrying out spreading in Humberside should furnish to this authority the particulars detailed above:

(i)   in a case where there is to be a single spreading, in advance of carrying out the spreading; or
(ii)   in a case where there is to be regular or frequent spreading of waste of a similar composition, every six months or, where the waste to be spread is of a description different to that last notified, in advance of carrying out the spreading.

Please send completed form to:

Example B: Adapted from Cornwall County Council (1995)

WRA Ref No.

WASTE REGULATION AUTHORITY

LAND SPREADING OF WASTE - NOTIFICATION OF DEPOSIT

Notification made under Regulation 17 of the Waste Management Licensing Regulations 1994

1. **FREQUENCY OF DEPOSIT**   (Tick Box)

   Occasional Deposit   [   ]

   Proposed deposit (date)

   6 monthly deposit   [   ]

   For the period   from (date)   to (date)

2. **DETAILS OF SPREADER (Person or company making deposit)**
   Name:

   Address: Post Code:

   Telephone Number: Fax Number:

3. **STORAGE**
   Where will the waste be stored prior to spreading?

   If the waste is to be stored on the land where it is to be spread, then the location of the storage area or lagoon

   should be shown on the field plan.

4. **DETAILS OF WASTE**   **ONLY ONE WASTE PER FORM**

   (a)   **Waste type**

         Indicate which type of waste is to be spread   (Tick box)

         Part 1
   a   Waste soil or compost   [   ]

   b   Waste wood, bark or other plant matter   [   ]

         Part 2
   g   Dredgings from any inland waters   [   ]

   (b)   **Producer** (except in the case of domestic septic tanks)

         Name:

         Address:

         Post code: Telephone Number:

   (c)   **Description**

   i   Process from which waste originated:

   ii   General description and physical nature of the waste:

   iii   Chemical analysis of waste (except for septic tank sludge):   Yes/No/Previously Supplied*

   iv   Microbiological analysis of waste (except for septic tank sludge):   Yes/No/Previously Supplied*

5. **QUANTITY**
   Quantity to be spread (if single load)   Tonnes   *or*

   Total quantity of this waste to be spread over next six month period   Tonnes

6. **DETAILS OF SITE**
   (In the case of large scale landspreading operations more substantial information may be required in the form
   of a Farm Waste Management Plan for the site, prepared by a suitably qualified person.)

   Landowner/Ocupier:
   Address:

   Grid Reference:

Continued

Box 6.6.2, Example B continued

7. **Does a Farm Waste Management Plan exist for site?**   *Yes/No
    (If **No** you **MUST** complete Sections 8 to 13 below)   (If **Yes** Attach Copy)

8. **Field Plan attached**   *Yes/No/Previously supplied

    Mark position on field plan of: watercourses, wells, ponds, boreholes, field drains, houses within 500 metres of spreading, etc.

    Indicate spreading areas.

9. **FIELD INFORMATION**

    O/S Field No.   Spreading area (Ha)   Crop   Special requirements

10. **SOIL INFORMATION**

    Soil type/depth   Subsoil type/depth   Soil analysis

11. **APPLICATION** (Method)

    Surface Spreader   Yes/No*
    Direct Injection   Yes/No*
    Rain Gun   Yes/No*
    Fixed Irrigation System   Yes/No*

    Rate: gallons/litres/tonnes per hectare*

    Frequency of Spreading: daily/weekly/monthly*

12. **OTHER WASTES APPLIED TO LAND WITHIN LAST 12 MONTHS**

    Description:
    Quantities: tonnes per hectare

13. <u>**DECLARATION**</u>

    Person who is to undertake the waste spreading
    I declare that the information provided is true to the best of my knowledge and that the waste is to be deposited in a manner which complies with current codes of practice so as not to cause the pollution of water, harm to health or to cause detriment to local amenities.

    Signed   ....
    Position   ....
    Date   ....

    Person who occupies the land
    I declare that I give permission for the spreading of the above shown wastes on land which I occupy.

    Signed   ....
    Position   ....
    Date   ....

    Activity of benefit to agriculture - qualified person
    I declare that on the basis of the information provided above, and subject to adherence to all Codes of Good Agricultural Practice, that the exempt activity concerned will result in Benefit to Agriculture or Ecological Improvement.

    Signed   ....
    Qualifications   ....
    Date   ....

    * *delete as applicable.*
    Return completed form to:

6.6.3 Land reclamation or improvement (change of use)

This practice includes spreading, carried out in accordance with any planning permission requirements, for the reclamation or improvement of land resulting in benefit to agriculture or ecological improvement, provided that by reason of industrial or other development the land is incapable of beneficial use without treatment. The maximum limit is 20,000 cubic metres per hectare (see Section 3.6).

Question	Answer
If the dredged material is to be disposed of for this purpose, how is it determined whether or not the disposal operation is exempt?	In many ways the same procedure as that for spreading onto agricultural land applies. The difference with this exemption lies in the quantities applied to the land and the current and subsequent use of the land. The land improvement operation may need planning permission (see Section 3.6) since the work involves a change of use.
Question	Answer
How will the quantities mentioned be measured?	The consultations we have undertaken have produced different responses. At the moment there is no clear answer, but it is most likely that measurement will be determined immediately the material has been placed by calculating the volume of dredgings on the land after spreading but before drying or compaction. An estimate of the volume after placing and spreading should be made at the planning stage when the physical characteristics of the material to be dredged are known. If this is not possible the bulk volume transported to disposal should be monitored and control exercised on these figures.
Question	Answer
Will quantities be measured for every operation?	The WRA visits the works occasionally so is aware of what is going on. If it is felt that too much material is placed, steps will be taken to assess the actual quantity put on site. However, under Duty of Care a record should be made of any quantities of wastes transferred.
Question	Answer
Who is responsible if too much is placed?	The Regulations are written in such a manner that, whatever the reason, the applicant will be held responsible.

6.6.4 Bank disposal

This practice enables the operator to dispose of dredged material along banks or towpaths, with a maximum disposal of 50 tonnes per day per metre of bank or towpath along which it is deposited (see Section 3.7).

Question	Answer
How do you comply with this exemption?	Again the activity needs to be registered with the WRA.
Question	Answer
Will the WRA want to know anything else?	Apart from registering the activity with the appropriate details, as mentioned in Section 3.4, no. But the responsibility to ensure that no pollution is caused to the waterways lies with the applicant. The NRA will closely monitor this aspect of disposal. It is therefore prudent for the applicant to understand the quality of the material they intend to dredge (see section 4.4).

6.6.5 Composting

When dredged material contains a large quantity of biodegradable matter, composting may take place at the location where the waste is produced or where the compost is to be used, or at any other place occupied by the person producing the waste or using the compost. This applies if the total quantity does not exceed 1000 cubic metres at any time (see Section 3.8.3).

Question	Answer
When composting of the dredged material has taken place, can the compost be spread on to land?	Yes, if composting has resulted in materials which may be spread on land for the benefit of agriculture or ecological improvement.

6.7 DISPOSAL UNDER LICENCE

Under the Waste Management Licensing Regulations all disposal of dredged material not qualifying for an exemption needs to be licensed. Several disposal options are available and are discussed in this section.

In addition to the form of disposal to be used, options can be distinguished based on who the owner of the site is and whether the site exists or is to be newly licensed. In a situation where the licensed site is owned by somebody else (i.e. not by the operator responsible for dredging the waterway) the framework is relatively simple (see Figure 2.9.4). This section discusses, *inter alia*, the advantages and disadvantages of each form of disposal.

6.7.1 Landfill sites

Landfill sites and lagoons are simply different forms of <u>licensed sites</u>. Specifically, a landfill is an area of agricultural or industrial land containing waste. The Waste Management Licensing Regulations will impose conditions concerning the precautions that are to be put in place to ensure that waste cannot leak from the site and to ensure that the public cannot be placed in danger by gaining access to the waste. New sites will require a design that strictly controls the liquid levels within them. Many of the larger existing sites are owned by multi-site operators, often where the filling of gravel workings or similar activities is taking place. Landfill sites tend to provide a permanent disposal site.

6.7.2 Lagoons

A lagoon is a licensed site for waste which is liquid when deposited. It normally constitutes an area of land which is enclosed by earth bunds or other containing structures on all sides to store waste. It may be used either as a permanent place of disposal or as a temporary storage or transfer station.

A lagoon can vary in size from very small to a significant area. All reasonable precautions must be taken to ensure that waste cannot escape from the lagoon and that members of the public are unable to gain access to the waste. When a lagoon is (to be) used in the process of disposing of dredged material, other than in the exempt activity of storage of dredgings for spreading on agricultural land (see Section 3.5.6), it will require a licence. In this case the precautions required above will be formalised as part of the conditions of the Waste Management Licence. The types of lagoons currently in existence vary considerably. For example, some lagoons cover approximately 80 hectares and can be built up to approximately 8 to 9 metres high. Other, smaller, lagoons are built for dewatering dredged material. The size of these lagoons can be as small as 50 square metres and approximately 1 metre high.

The technical feasibility and environmental considerations associated with using <u>licensed sites</u> are considered in Table 6.7.2(a).

Table 6.7.2(a) Licensed sites: landfills and lagoons

Option	Technical feasibility	Environmental considerations	Legislative constraints
Lagoon or landfill site	- In theory there are no technical constraints placed upon designing a disposal site. However, the technology applied may turn out to be very expensive when, for instance, the local geology shows a permeable stratum and a number of protective measures are required by the authorities to prevent any contamination or leaching into the surrounding environment. Such Regulations and measures are described in Waste Management Papers 4, 26, 26A and 27. Measures required might include: gas monitoring points, protective liners, underlying drainage, etc. - In addition, the bunds surrounding the planned sites must be appropriately designed, preferably by a third party to the dredging operation (such as an experienced and qualified geotechnical engineer). This approach is likely to receive the approval of the HSE, who are a statutory consultee in the licence application procedure.	- The main issues in respect to a disposal site will generally be the impact on the local environment. Leaching of contaminants, methane production, wind-blown dusts and noise by operating machinery are all examples of environmental parameters that should be carefully investigated when disposing of dredged sediment. Many of these parameters will already have restrictions placed upon them. - Each Waste Management Licence will have different requirements depending on the size and location of the site and the requirements of the WRA. Specific requirements may concern health and safety aspects (e.g. complete fencing around the site; site access to be restricted when the site is located in an area with more than one owner, etc.). - Consideration should be given to whether the site might be converted, when disposal is completed, into a facility such as a golf course, amenity park or other beneficial use. - Surface water and groundwater quality considerations.	- When disposing to an existing licensed site owned by someone other than the owner of the dredged material, the legislative framework is very straightforward (see Figure 2.5.2). - Disposal to such sites requires notification from the owner of the dredged material to the owner or person in charge of that particular disposal site. Quantities and quality generally need to be confirmed as the basis on which the price of disposal can be decided. - When material is disposed of at a licensed site, the responsibilities set out in the Duty of Care will be assumed by the licence holder, provided that the transfer complies with the legal requirements. The Duty of Care requires that holders of waste must take all reasonable steps to prevent its escape or unauthorised handling, and only transfer it with proper documentation to an authorised person (see Section 3.11). - Paperwork recording the above needs to be kept for 2 years for possible future inspection by the WRA.

The advantages and disadvantages of an existing licensed site versus a site to be newly licensed by the owner of the waste, however, are different and these are summarised in Table 6.7.2(b).

6.7.3 Disposal at sea

For those operators with access to tidal waters, one possible alternative to the disposal of dredged material on land may be disposal at sea. Most of the dredged sediment which is currently disposed of at sea originates from the capital or maintenance dredging of ports, harbours, marinas, etc. Once material has been dredged, it is generally transported to a specified disposal site designated by MAFF. Licensed sites for marine disposal are not necessarily in offshore, high-energy environments but may be located in estuaries (e.g. Salcombe and Kingsbridge Estuary). These disposal sites were originally designated in response to specific needs, for example, those associated with port operation. However, applications for new sites are unlikely to be successful unless it can be demonstrated that no other (beneficial) use can be found for the dredged material (such as land reclamation, construction of flood defence, etc.). In considering applications for new marine disposal sites, the full investigation required for all other possible disposal options (including land disposal) can be both costly and time consuming.

The technical and environmental considerations and advantages and disadvantages of the option of sea disposal are given in Table 6.7.3.

Table 6.7.2(b) Lagoons and landfill sites - advantages and disadvantages

Option	Advantages	Disadvantages
Existing licensed lagoon or landfill site	- Since all the responsibilities set out in the Waste Management Licence are with the licence holder, there are no further costs involved for the owner of the dredged material. Costs associated with monitoring, licence fees, etc. (until the licence is surrendered) are therefore not relevant unless the disposal site is also owned by the dredging operator. The only costs for the disposer are the transport to the site and charges for disposal by the site owner.	- Depending on its classification, the material will be charged differently (see Section 3.13). Each class of dredged material will fall into a different charging band. This means for instance that material which is classified as contaminated is more expensive to dispose of than material which is less contaminated. - The high moisture content of the material may increase costs significantly. - The disposal site may not necessarily be near the location where the dredging takes place and therefore a cost-benefit analysis, comparing the cost of transport and disposal costs on a site owned by a third party against any costs incurred when the owner of the dredged material can dispose of it on a site owned by him/herself, should be carried out. - There may be uncertainties as to the volume which can be disposed of on such a site.
New licensed lagoon or landfill site	- The location for a lagoon or landfill site can be planned in advance and can, therefore, be identified to reduce transport requirements and costs. For instance, in an industrial area where the sediments to be dredged are likely to be contaminated, an exemption is unlikely (unless a buyer is found for the material). Local disposal may therefore be the best and cheapest option. - Even if a licence is applied for a lagoon or landfill for permanent disposal, a beneficial end use may eventually be identified and cost savings achieved. When the market is right and a buyer is found, the material may be sold with modification to the licence. - After storage (in a lagoon) the material will have dried to a stage in which it can be transported more easily and it may be a viable engineering use (see Section 6.3). - A site established as a waste transfer station can be used for a number of dredging operations in the same geographical area subject to the conditions laid down on monitoring and quantities (see Section 6.2.3).	- The application procedure for obtaining licences for newly planned sites need to be started well before dredging is undertaken. Some new licences may require completion of a full Environmental Assessment, Working Plan, Waste Management Application Form and Procedure for Construction of Dredging Lagoons. This may involve long periods of time before the planning authorities and the WRA are satisfied. - With licensed sites, costs will be incurred for fees and charges, for design, preparing a Working Plan, and for any required monitoring scheme (the latter may also include a stringent sampling scheme). - Technically competent landfill manager(s) will need to be appointed for the site(s). - Obtaining a licence may be very difficult, depending on the approach adopted by the local WRA(s) and on the nature of the disposed dredgings. Long-term monitoring might be required during which fees and charges are paid before a licence is accepted for surrender. - The WRA may impose other requirements on a site such as dust control, early seeding, fencing, etc. - If a licence is refused by the WRA, an appeal can be made to the Secretary of State within 6 months of the date of refusal. However, in practice this could mean that, after all the work that is carried out, (including obtaining planning permission) there is no certainty that a Waste Management Licence will be issued. The grounds on which a licence may be refused are pollution of the environment, threat to human health and serious detriment to the amenities of the locality. - The intended volumes for the site may not be agreed upon by either the planning authorities or the WRAs (i.e. restriction in height, size, etc.). - The authorised volume for the site may not justify the development investment needed.

Table 6.7.3 Sea disposal

Option	Technical feasibility	Environmental considerations	Advantages	Disadvantages
Sea disposal	- The suitability of sediment-carrying river barges for marine disposal of dredged material should be considered, although double handling would remove this constraint.	- Physical smothering of marine organisms on the sea bed at or adjacent to the disposal site. - Remobilisation of pollutants, especially if exposed to oxygenated water which may allow uptake of contaminants by marine organisms. - Increased turbidity in and around the disposal site may adversely affect marine communities (e.g. by reduction of photosynthesis). - Interference of vessels carrying dredged material with other users of the sea (e.g. fishermen).	- If permissions can be obtained, the sites can frequently receive large quantities of material. - May prove relatively cost-effective.	- May sometimes require a lengthy licensing procedure (Food and Environment Protection Act 1985; see also Section 3.17). - Increasing congestion and conflict in coastal and estuarine areas may lead to operating constraints. - Distance of transport of material to disposal site (although this is obviously dependent on the dredging location). - Monitoring of a marine disposal site, which may be required in the licence conditions, may be expensive and technically problematic. - Controls on accuracy of the disposal operation are stringent and would need to be developed for inland craft, or double handling may be required. - MAFF is likely to require a great deal of work to be carried out to show that marine disposal is the only reasonable option.

7 Health and safety

7.1 INTRODUCTION

The storage, handling and disposal of dredged material (Chapter 6) potentially poses a number of risks to workers' health and safety, whether or not the material is regarded as hazardous. Those dealing with dredged material should be aware of the relevant legislation and in particular the Health and Safety at Work etc. Act 1974. Other relevant legislation includes:

- Management of Health and Safety at Work Regulations 1992
- Control of Substances Hazardous to Health Regulations 1994
- Construction (Design and Management) Regulations 1994
- Personal Protective Equipment at Work Regulations 1992.

In addition to these general requirements, a number of specific concerns were highlighted by the HSE during consultation. The main statutory requirements are discussed in Section 7.2.

7.2 STATUTORY REQUIREMENTS

7.2.1 Health and Safety at Work etc. Act 1974

All employers in the United Kingdom have to comply with the Health and Safety at Work etc. Act 1974 (the Act). The Act consists of four parts. Part 1 is the part of the Act concerned with Employer's General Duties and contains provisions for:

- the health and safety of people at work
- protection of others against health and safety risks from work activities
- control of danger from articles and substances used at work
- controlling certain atmospheric emissions.

The General Duties of the Act require both the employer and the contractor to ensure the health, safety and welfare of not only their own employees but also the employees of others (including the general public).

Key responsibilities and duties defined in the Act are summarised in Box 7.2.1. However, this list is not comprehensive and reference to the Act should be made for a full appreciation of the duties of employers.

7.2.2 Construction (Design and Management) Regulations 1994

The Construction (Design and Management) Regulations 1994 (CDM Regulations) were introduced on 31 March 1995 and came fully into force on 1 January 1996. The definition of "*construction work*" given in the CDM Regulations will apply to most dredging projects.

It is a duty of employers to ensure that those they appoint are competent and will allocate adequate resources for health and safety. The CDM Regulations also require the employer to appoint a planning supervisor, to coordinate the healthy and safety aspects of the project design and initial planning, and a principal contractor, to take over and develop the health and safety management and to coordinate the activities of all contractors involved with the project.

Box 7.2.1   Health and Safety at Work etc. Act 1974

Section 2(1)	It is the duty of every employer to ensure, so far as is reasonably practicable, the health, safety and welfare at work of all his employees.
Section 2(3)	It is the duty of every employer who employs five or more employees to prepare and, as often as may be appropriate, revise a written statement of his general policy with respect to the health and safety of all employees and the organisation and arrangements for the time being in force for carrying out that policy.
Section 3(1)	It shall be the duty of every employer to conduct his undertaking in such a way as to ensure, so far as is reasonably practicable, that persons not in his employment who may be affected thereby are not exposed to risks to their health or safety.
Section 3(2)	It shall be the duty of every self-employed person to conduct his undertaking in such a way as to ensure, so far as is reasonably practicable, that he and other persons (not being his employees) who may be affected thereby are not thereby exposed to risks to their health and safety.
Section 4(2)	It shall be the duty of each person who has, to any extent, control of premises to which this section applies or of the means of access thereto or egress therefrom, or of any plant or substance in such premises, to take such measures as it is reasonable for a person in his position to take, to ensure, so far as is reasonably practicable, that the premises, all means of access thereto or egress therefrom, available for use by persons using the premises, and any plant or substance in the premises, or as the case may be, provided for use, is or are safe and without risks to health.
Section 4(3)	Where a person has by virtue of any contract an obligation of any extent in relation to: (a) the maintenance or repair of any premises to which this section applies or any means of access thereto or egress therefrom; or (b) the safety of or the absence of risks to health arising from plant or substances in any such premises; that person shall be treated, for the purposes of Section 4(2) above, as being a person who has control of the matters to which his obligation extends.
Section 6(3)	It shall be the duty of any person who erects or installs any article for use at work, in any premises where that article is to be used by persons at work, to ensure, so far as is reasonably practicable, that nothing about the way in which it is erected or installed makes it unsafe or a risk to health when properly used.

Dredging contractors should expect to appoint a principal contractor. They will be required to demonstrate their competence and to demonstrate that they will allocate sufficient resources for health and safety, perhaps as part of a prequalification exercise. Typical areas of inquiry may include the dredging contractor's health and safety policy, accident record, any actions taken by enforcing authorities, quality assurance procedures, training policy and evidence of technical competence.

Part of the planning supervisor's duties are to ensure that a pre-tender health and safety plan is prepared and this should be included with any tender documentation. The principal contractor appointed is required to develop the plan into a construction phase health and safety plan, which will form the foundation upon which the health and safety management of the construction work will be based. A construction phase plan is expected to include arrangements for ensuring the health and safety of all who may be affected by the construction work and for the management of health and safety during the construction work. The employer has a duty under the CDM Regulations to ensure the construction phase of a project does not start until an adequate construction phase health and safety plan has been prepared by the principal contractor.

7.2.3 Statutory obligations

Contractors (operators) are expected to be aware of all current Acts, Regulations and Orders applicable to their work activities. They must ensure that they work in accordance with this legislation as failure to do so may result in work being stopped, exclusion from the site, or prosecution.

7.2.4 Safety policy statements

As required by Section 2(3) of the Health and Safety at Work etc. Act 1974, any contractor (operator) who employs five or more persons shall have a written Safety Policy Statement. Main contractors should see copies of all subcontractors' Safety Policy Statements before they commence work. The CDM Regulations require main contractors to submit a copy of their Safety Policy Statement to the project manager with tender documentation.

7.2.5 Method statements

Every contractor must carry out an assessment of the risks of undertaking work on the employer's premises in accordance with the Management of Health and Safety at Work Regulations 1992.

A Method Statement is a detailed account of how a job of work must be carried out and the following guidelines should be applied.

(a)   Full details should be provided of:

*   the work to be undertaken and method
*   the purpose of the job
*   the exact location of the job and means of access
*   the expected starting date and duration of the job.

(b)   The staffing arrangements, including nomination of a Person in Charge, should be included.

7.2.6 Personal protection

It should be noted that it is a requirement of the Management of Health and Safety at Work Regulations 1992, the CDM Regulations 1994, and the Control of Substances Hazardous to Health (COSHH) Regulations 1994 that the issue of personal protective equipment should only be done as the final act in the "hierarchy of risk control", when other measures to protect employees are not reasonably practicable. In every case the first action to be taken is the avoidance of the hazard. If this is not reasonably practicable, then steps should be taken to minimise the risk to all persons by design or engineering means. Only then, if it is not reasonably practicable to do so by such means, should consideration be given to the issue of personal protective equipment.

Contractors (operators) must ensure that suitable personal protection is provided on site, for use by their employees, subcontractors and other persons who need to visit the site or building operations, in accordance with all relevant Regulations, Acts, and approved Standards. Contractors (operators) must ensure that their operatives are trained in the selection, use, maintenance and storage of all personal protective equipment. It is equally important that operatives are made aware of the personal protection requirements particular to working in a water environment and disposing of sludge in lagoons and/or on banks.

In particular, personal protection must, where necessary, include life jackets, safety helmets, goggles, respiratory protection, hearing protection, head protection and clothing (e.g. disposable overalls). Suitable footwear to minimise the risks from slippery sludges (e.g. barges, quayside) should always be worn. Footwear would also need to be chemical- and oil-resistant and be reinforced to relevant British Standards.

Safety helmets must be worn by all persons on sites or operations where there is a risk of head injury. It is the responsibility of the main contractor to ensure that suitable head protection for all staff is provided and worn. Suitable warning signs must be displayed in all areas designated by the main contractor to be "Head Protection Areas". Appropriate warning signs must be displayed in any areas designated as "Head", "Eye" or "Hearing Protection Zones". The provision of "Head Protection Areas" signs is particularly important in areas where the public may have a right of way or in cases where the working sites are next to or within urbanised areas.

Dealing with dredged material, particularly from lakes or static waterways, can lead to exposure to Weil's disease, also known as *leptospirosis*. Bacteria of the genus *Leptospira*, which can enter the human body via abrasions on the skin and through the mucous membrane, can cause acute kidney, liver and central nervous system infections. Each employee involved in dredging and disposal operations should be formally issued with details of the disease and steps to be taken in the event of a suspected condition. Operatives should be asked to advise their GP of their working environment so that the correct treatment is initiated in the unlikely event of Weil's disease being contracted. Personal protection in dredging/disposal projects therefore should include gloves, face masks (to protect against splashed water) and waterproof clothing (including boots), particularly for those operatives working on barges and at the sorting station. The sorting station should also be provided with washing facilities (see below).

More often than not, the dredged material from urban waterways contains large debris which has to be removed prior to either transferring to a licensed disposal site or disposing onto banks or agricultural land. Such debris can comprise sharp objects and/or containers with hazardous substances (e.g. drums containing skin irritants or flammable substances). All operatives involved in dredging/sorting prior to disposal projects should be made aware of all physical, chemical or microbiological hazards they might encounter; of all necessary measures to eliminate exposure to those hazards; and of all measures to reduce the intensity of a hazard should an event occur.

7.2.7 First aid and site facilities

Contractors (operators) must ensure that adequate First Aid Provision is made for persons working at their site(s) on the employer's premises. However, in the event of an emergency the employer's Occupational Health Staff, where available, may be called on for assistance. The contractor responsible for providing first aid must ensure that all other contractors working at the site are aware of the procedures to be followed. For isolated sites, a first-aid kit must be made available on site. A mobile phone is advisable and may be required by licence conditions.

The Contractors (operators) must ensure that both washing facilities and fire fighting equipment are made available at the working site(s) and that they are properly serviced and maintained. Washing and fire fighting equipment should be serviced regularly and certificates of service should be on display.

Where lagoons or areas of wet sludge or dredged material are located, the public should be excluded by means of secure fencing and warning signs erected as to the dangers within (e.g. "Danger - Deep Water"). Life-rings and/or throw-lines should be provided at regular intervals for emergency use.

7.2.8 Notification of intended works

It is the contractor's (operator's) responsibility to notify the HSE of any disposal operation or work of engineering construction as specified in current legislation.

7.2.9 COSHH assessments

Before any disposal (or sorting/treating) operation begins, an assessment of the possible hazards likely to be encountered must be undertaken in accordance with the Control of Substances Hazardous to Health (COSHH) Regulations 1994. Although personal hygiene is one of the most important protective measures against chemical and pathogenic hazards in particular, all operatives at the working site, including subcontractors, should be covered by a COSHH assessment. COSHH implementation involves a risk assessment and control exercise of the risk(s) involved in dredging/sorting/disposal of dredged material operations. It is a statutory requirement that a COSHH assessment should be undertaken before any work is carried on a site involving any one of the aforementioned three operations which are liable to expose operatives to a substance hazardous to health. A COSHH assessment of dredged sediment should fulfil the following requirements:

- evaluation of the risks to health arising from the nature of the work and handled material
- identification of control measures in accordance with the COSHH Regulations 1994.

Evaluation of the risk involves three parameters which need to be site, activity or material specific. These three parameters are:

- context
- hazard
- exposure.

For detailed information on what a COSHH assessment should entail and the range of chemical and pathogenic substances hazardous to human health the reader should refer to CIRIA's *Guide to the Control of Substances Hazardous to Health in Design and Construction* Report 125 (CIRIA, 1993a), CIRIA's *Guide to Safe-Working Practices for Contaminated Sites* (CIRIA, 1993b), as well as SI 1994 No. 3246 and HSE's reports on hazardous substances.

7.2.10 Confined spaces

As part of and in addition to the COSHH assessments, all sites and equipment connected with dredging and disposal activities should be assessed to identify confined spaces. Typical examples may be hoppers of barges, containers of sludge or tunnels. Each confined space should be clearly described and rules laid down to control the entry to such locations including method, protective clothing/equipment required, and notification procedures.

7.3 HEALTH AND SAFETY EXECUTIVE CONCERNS

7.3.1 Background

The following section details the solicited concerns of the HSE. The HSE is particularly concerned that the storage of dredged material above ground, where that material is a liquid or sludge or in any way has the potential to flow, is done so in a manner which will ensure stability of the store. This requires the application of sound geotechnical engineering principles to the design, construction and monitoring of the store. These activities should be carried out under the control of a competent geotechnical engineer. Routine monitoring of the store for pollution due to leachate or runoff in no way compensates for any lack of geotechnical monitoring. Consideration should be given to the effect that biodegradation of the dredged material may have on stability.

7.3.2 Suggested good practice

The HSE is eager to ensure the following good practice:

- design of the lagoon by a competent geotechnical engineer
- construction and filling of the lagoon under the supervision of a competent geotechnical engineer
- geotechnical investigation of the site of any disposal lagoon prior to disposal operations
- adherence to a scheme of regular inspection of the lagoon by a competent geotechnical engineer and a system for remedying any faults detected during the inspection within an appropriate timescale.

7.3.3 Consultation between HSE and WRA

HSE is a consultee of all WRAs on matters relating to hazardous activities. Discussion with various WRAs, in particular the West Midlands Hazardous Waste Unit, has revealed that when an application for a disposal site for dredged sediment is processed, the HSE will be contacted about the stability of the confined site as well as the general safety of those at risk. The disposal of dredged sediment in brick pits and natural depressions does not appear to give any particular reasons for concern to the HSE, unless there is a need to build a bund to increase a lagoon's capacity.

Normally the WRA will resolve the type of issues raised in this section, possibly through consultation with the HSE. When a licence application is made, the WRA may consult the HSE, who will then have 21 days to respond as part of the 4-month total consultation period. Two areas of specific concern to the HSE are outlined below.

7.3.4 Working plans

The applicant should, therefore, take due care that the working plans submitted to WRA with the Waste Management Licence application will satisfy the HSE's requirements since the HSE is a statutory consultee under the Regulations. A prospective applicant would benefit by contacting the two regulators in order to identify their requirements in advance and hence speed up the procedure.

7.3.5 Spreading on agricultural land

HSE Agricultural Inspectors may come across the spreading of dredged material in the course of their normal duties. Should an Agricultural Inspector come across a spreading operation which gives rise to concern he or she will at that time take whatever action he or she considers appropriate, including seeking the advice of specialist civil engineering inspectors. Spreading operations *per se*, do not have to be notified to HSE.

8 Financial aspects of dredging disposal

8.1 INTRODUCTION

The costs associated with the disposal of dredged material will be dependent on the nature of the material and the scale of the disposal operation. In general, larger operations will probably incur lower per unit costs due to economies of scale.

There are some cost components which are likely to be common to many disposal options, where they relate to the analysis, treatment, storage, or transport of dredged material. These costs are discussed in Section 8.2.

The costs associated with the disposal of dredged material have been examined in relation to the following disposal or use options:

- disposal to agricultural land
- disposal for reclamation or improvement
- bankside disposal
- disposal to a lagoon or landfill
- sea disposal
- sale of materials for re-use.

The ranges of possible cost components associated with each disposal option are discussed in Sections 8.3 to 8.8. Section 8.9 provides a summary table which indicates the different stages where costs could be incurred under the different options.

To aid the discussion of different disposal options, estimates of possible cost components (1993/94 prices) have been included where available. It should be noted that these costs are order-of-magnitude estimates for guidance only, and that the actual costs incurred will be highly site specific and could thus vary from those given. Where possible, site specific "real" costs have been included in the form of case studies.

8.2 COSTS ASSOCIATED WITH ANALYSIS, TREATMENT, STORAGE AND TRANSPORT

The type of costs associated with the analysis, treatment, storage and transport of dredged material could be common to many disposal options. Whether or not such costs are actually incurred will depend on the nature of the dredged material and the disposal options which are available. In some cases, for example, transport may be necessary if there is no disposal site adjacent to the dredged area.

8.2.1 Analysis of materials for disposal

Analysis of materials is likely to be required to prove the quality of the material. Costs will vary according to the number of samples, how many substances are to be tested for, the nature of those substances and the frequency of testing. The exact nature of sampling and testing requirements will vary according to the nature of the dredged area and the disposal route. The need for such sampling and analysis (if any) should be discussed with the WRA or NRA at the earliest possible opportunity. Tests for the following may be required: metals content (£5 per metal); total organic matter (£15); pH (£3); volatile organics (£20 to £100); PAHs and phenols (£50 to £60); pesticides and herbicides (£90); PCBs (£50 to £300); and NRA leach tests (£15). It should be noted that the above costs are indicative and could alter substantially from laboratory to laboratory.

In addition to testing costs, there will be sampling costs. These will vary according to who carries out the sampling (e.g. the testing laboratory or the owner of the dredged material) and their associated travel or transportation expenses, equipment hire, etc.

8.2.2 Treatment and storage

Treatment of dredged material can be undertaken to dewater it, to separate different grain sizes, and/or to remove contamination (see Section 6.4).

Dewatering can be used as a means of reducing transport and hence disposal costs, as volumes can potentially be reduced by up to a factor of three (for material with a high (e.g. 70%) water content). Dewatering can be achieved naturally as a result of lagoon storage (with the associated costs set out in Section 8.6) or by mechanical dewatering.

A Waste Management Licence is required for all forms of treatment and the associated fees are set according to the Waste Management Licensing (Fees and Charges) Scheme 1995. However, where materials are to be stored on-site prior to an exempt use and without further treatment (see Sections 8.3 to 8.5), then such storage does not require a licence and will not incur the associated costs. In certain circumstances (e.g. where the storage may affect drainage or flood alleviation) a consent from NRA may be required.

Studies carried out by British Waterways have indicated that treatment costs could vary from around £11/m^3 for dewatering only to in excess of £80/m^3 for full treatment to reduce contamination.

8.2.3 Transport

Transport costs will vary according to the mode of transport (road or water), the distance travelled and the water content of the material. Conventional road lorries cannot be loaded to capacity with wet dredgings. This consequently increases costs and could make the use of specially prepared lorries or dewatering treatment more cost-effective options.

British Waterways have found that, for road transport in fully loaded 15 m^3 capacity lorries, costs (including loading and unloading) can be as high as £11/km for distances less than 15 km. Average haulage costs are of the order of £0.25/m^3 per kilometre, with some materials requiring transport over 100 km due to lack of suitable disposal sites.

8.3 DISPOSAL TO AGRICULTURAL LAND

The costs associated with the (exempt) disposal of dredged material to agricultural land (see Sections 3.5 and 5.5) could consist of the following components (case study examples are presented in Box 8.3.1):

- the cost associated with identifying suitable land and any negotiations
- any charge made by the farmer for the use of land (or compensation payment)
- any transport, treatment and storage costs
- spreading costs
- the costs associated with obtaining an exemption
- other costs such as those associated with requirements for fencing and restoration.

The NRA does not have statutory powers in this area and thus there is no NRA licence fee. Should spreading result in water pollution, however, the NRA can prosecute and this could result in legal costs (court costs and associated fines) as well as costs directly related to any clean-up of pollution deemed necessary. It should be noted that, in such cases, legal costs often exceed clean-up costs.

Operators contacted during the consultation process for this document indicated that the costs associated with compensation payments to farmers (for example, for temporary loss of land use) can vary from nothing to £5000/ha per disposal. At present, a large number of agricultural disposal operations incur no costs. However, it should be noted that some of those involved consider that such costs could increase as the awareness of landowners increases with respect to the value of land for waste disposal. Indeed, some landowners are already seeking advice from land agents, etc. to ensure that they obtain the market value for the use of their land.

Box 8.3.1   Costs associated with disposal to agricultural land

---

Drainage Ditches: In some areas, the dredging of drainage ditches is carried out by the internal drainage boards (IDBs) and almost all material is disposed of to agricultural land adjacent to the waterway (i.e. there are no transport costs). Thus, the costs associated with the disposal operation are simply the cost of spreading the material. In most cases, the material is spread immediately after dredging, although materials are sometimes stored on the bankside and spread later to avoid crop damage (the increased cost of disposal being less than the costs associated with compensation for damage to high-value crops such as celery and lettuce). The Association of Drainage Authorities considers that there will be minimal additional costs associated with the 1994 Waste Management Licensing Regulations as the IDBs may seek blanket cover for all such dredging operations.

Barton Broad: The Broads Authority is planning to dredge 260,000 m^3 of sediment from Barton Broad and to dispose of it to agricultural land. This will involve lifting the topsoil from fields to form lagoon bunds (and topsoil stores). When sufficient dredged material has been placed in the lagoons, the topsoil will be spread back over the dredgings and cultivated. In investigating the feasibility of this disposal option, the Broads Authority has undertaken a series of surveys, etc. and has incurred the following costs: bathymetric surveys (£5,000); chemical analyses of sediment (£5000); density profiling of sediment (£5000); soil survey, agricultural analysis of lagoon site and agricultural advice (£5000); dredging trial (£5000); feasibility study (£15,000); rental of ancillary land and compensation for shooting rights (£10,000); and preparation and analysis of tender documents for dredging and disposal contracts (£8000). The total costs associated with setting up this disposal route are thus in the order of £60,000.

The costs associated with obtaining an exemption alone could range from virtually nothing to tens of thousands of pounds depending on the nature and quantity of the dredged material and thus the work required. Costs could be greatest where the suitability of the dredged material is under question (e.g. where there is concern about contamination because of the nature of the area from which the sediment is dredged). In such cases, the WRA may require detailed analyses of dredged material (e.g. heavy metals content, etc; see Section 4.4), information on the agricultural land and its environs, and a report from a competent person (showing that there will be agricultural benefit, indicating appropriate application rates, containing a pollution risk assessment, etc.). In all cases, early consultation with the WRA is recommended as a means of attempting to minimise the costs associated with obtaining an exemption. The operator can propose and agree with the WRA the metals to be analysed for and any other requirements.

ADAS have indicated that a detailed assessment of the suitability, etc. of agricultural land for the disposal of dredged sediment would involve the use of a soil scientist and would cost in the region of £1000 upwards. An assessment of the dredged material itself could cost anything in the region of £500 to £10,000 depending on the nature of the tasks to be undertaken (i.e. from a relatively simple analysis of the material to assess heavy metals content and conditioning effect to a full assessment including, for example, a year-long growth trial). Finally, the cost of hiring consultants to identify suitable land for disposal and to do everything required of them to obtain an exemption could cost in the region of £500 to £30,000 depending on the level of input necessary, which is in turn related to the nature and quantity of the dredged material.

8.4   DISPOSAL FOR RECLAMATION OR IMPROVEMENT

As for other exempt activities, the spreading of dredged material on land for reclamation or improvement (see Section 3.6) must be registered with the WRA. The costs associated with this disposal route may include:

- the cost associated with identifying suitable land and any negotiations
- any charge made for the use of land
- any transport, treatment, storage and spreading costs
- the costs associated with obtaining an exemption.

The costs associated with obtaining an exemption are likely to be the same as those for agricultural land, but may also include the cost of obtaining planning permission.

8.5   BANKSIDE DISPOSAL

The costs of (exempt) bankside disposal of dredged material (see Section 3.7) consist of any transport and/or storage costs and any costs associated with obtaining an exemption (possibly including sampling and analysis to satisfy the WRA that the proposed disposal meets the "relevant objectives" of the Waste Management Licensing Regulations and that the exemption applies). In addition, there may be costs associated with the need to relocate footpaths (alongside canals, for example). Where material is deposited on the bank where dredging took place, there will be no transport and storage costs. In the majority of cases, the costs associated with obtaining an exemption could be limited because the application procedure is likely to be simple (e.g. a letter to the WRA setting out the proposed dredging operations).

8.6   DISPOSAL TO A LAGOON OR LANDFILL

8.6.1   Introduction

The costs associated with disposal to a landfill or lagoon site (see Sections 6.7.1 and 6.7.2) will include the costs of disposal at a suitably licensed site, the costs of transport to that site and any prior treatment and/or analysis costs.

The costs of disposal to a commercial site can vary from less than £5/m^3 to around £90/m^3 depending on the quality and water content of the dredged material and the geographical location. As a result, in some cases, it may be cost-effective to undertake dewatering. For example, with disposal costs of £15/m^3 for material with 70% water content and £10/m^3 for material with 20% water content, dewatering would achieve an overall saving of £10/m^3 after volume reduction. In addition, dewatering may be a condition of disposal, as most commercial sites are limited by their licence conditions to a certain quantity of wet material.

It should be noted that any large debris resulting from the dredging process will probably need to be landfilled and should be considered separately from the bulk of dredged material (for which a number of disposal options may be available).

Whether or not it will be feasible and cost-effective to develop a site for the disposal of dredged material will depend on the quantities of materials to be disposed of, their nature, the cost of transport to an appropriate disposal facility and the cost of disposal at that site. The relatively high costs which can be associated with site design and obtaining planning permission could result in the costs of developing new landfills for dredging wastes being prohibitively high for smaller operations.

In developing a landfill or lagoon, consideration will need to be given to any costs associated with the following:

- land purchase or lease
- site design
- obtaining planning permission
- Waste Management Licensing Fees
- site construction
- operation - including the costs of staff (and any training), machinery, monitoring, licence fees, NRA fees for discharge consents, etc.

- closure (e.g. capping, any landscaping requirements)
- aftercare (e.g. monitoring)
- use of a consultant.

Some of these cost items are described in greater detail below, and case study examples of the costs associated with the development of lagoons are given in Box 8.6.1.

Box 8.6.1   Costs associated with the development of a lagoon or landfill

British Waterways has experience of developing a number of small disposal sites around the country with capacities in the order of a few thousand tonnes. Where regrading has been required, the associated costs have been in the order of £5000 to £20,000 per site. For removal of material to create a void, the costs have been of the order of £6/t. Interestingly, in the development of sites of this scale, one of the most expensive cost items has been the provision of hundreds of metres of security fencing at around £40/m. Other large items of expenditure have included land purchase and lease.

The Broads Authority has developed a lagoon for the disposal of material from the dredging of Hoveton Little Broad and Pound End. The lagoon cost around £10,000 to construct and is for the disposal of around 120,000 m³ of sediment.

Manchester Ship Canal Company is developing a new lagoon (No. 6 disposal ground) on its own land for the disposal of around 500,000 tpa of their dredged material. The lagoon covers an area of 70 ha and has around 8 million m³ of dredging space. Construction of the 9 m-high lagoon bunds required a 1 m-deep sand base (for which 180,000 m³ of sand was imported) and the on-site excavation of marshland to provide the other necessary materials. Costs associated with the control of the liquor resulting from dewatering will be around £25,000 per annum, plus £3000 per annum for NRA consent fees.

8.6.2 Fees for NRA discharge consents

The fees for NRA discharge consents are set according to a formula which is laid down in the NRA document *Proposed Scheme of Charges in Respect of Applications and Consents for Discharges to Controlled Waters* (NRA, 1991). This guidance is annually updated and presented in a NRA leaflet entitled *Annual Charges - Discharges to Controlled Waters*. There is an application fee and an annual charge, with the latter being set according to the volume and nature of the discharge and the nature of the receiving waters.

8.6.3 Fees for Waste Management Licensing

Under the Waste Management Licensing Regulations, the operators of a lagoon or landfill site must hold a Waste Management Licence. The associated fees are set according to the Waste Management Licensing (Fees and Charges) Scheme. Fees are payable for application, holding, modification and surrender. The most significant part of the charges relate to the size and type of site and the nature of material and rate of deposition. Clearly licence fees can be minimised by limiting the type and annual input of wastes to the site, although this must be balanced against any losses in potential revenue from the use of the site by others. There is no maximum fee or ceiling figure for fees for planning applications. The fees will depend on the situation in hand, for example, how much baseline data is collected, whether an appeal to the planning decision takes place, etc.

8.6.4 Construction costs

Construction costs will depend on the design of the site and will include costs associated with licence requirements (for example, fencing, lining and/or capping). Where it is considered that there is a risk to groundwater associated with the operation of a lagoon or landfill site, the NRA may require that the site is lined with clay and/or a flexible membrane liner (e.g. HDPE - see Section 5.4.4). It should be noted, however, that in the case of dredged material, such requirements are likely to be required only in extreme circumstances. The costs of using such protective materials include not only the cost of purchase (of the order of £5/tonne delivered for clay) but also the costs of placement. Such materials must be placed by experienced operators in a quality-controlled environment.

For materials which have a high organic matter content, gas collection and monitoring systems may be required. Where groundwater is present, groundwater monitoring boreholes may be required. As a result of licence requirements, British Waterways has found it necessary to install monitoring boreholes at some of its sites and has indicated that the cost of drilling a gas (or water) monitoring borehole is of the order of £150. Clearly, there will be recurrent costs associated with monitoring these boreholes which may continue after site closure as a result of any post-closure requirements.

8.6.5 Design and consultancy fees

Waste facilities should be designed by a qualified engineer with design costs varying according to the scale of the facility and the nature of the waste and surrounding area (i.e. the complexity of the design). In addition, it may be necessary to use consultants to obtain a facility's Waste Management Licence and/or to undertake any necessary environmental or risk assessment work. The costs associated with design and consultancy fees could vary from a few thousand pounds to in excess of £100,000.

8.7 SEA DISPOSAL

Under the Food and Environment Protection Act 1985, Part II, it is possible to dispose of any dredged material to sea provided a MAFF licence is held. The costs associated with such sea disposal will include:

- transport (and dumping) costs
- the cost of obtaining the licence
- the licence fee.

The costs associated with the transport and dumping of dredged material to sea will depend on the nature of the shipping used and the distance to the disposal site. Costs for the sea disposal of dredged material from the Harwich Haven Approach Channel are presented in Box 8.7.1.

In determining whether or not to grant a licence, MAFF consider a number of factors, including the nature and quantity of the dredged material and possible dumping sites, in order to determine the implications of the proposed scheme for the marine environment. In addition, MAFF has a duty to "*have regard to the practical availability of alternative methods of dealing with (the dredged material)*" and if it considers that (some or all of) the material could be alternatively or better utilised (e.g. used beneficially) then it will not grant a sea disposal licence for that material. The costs associated with obtaining a MAFF sea disposal licence are therefore the costs associated with proving that sea disposal is the only practicable option in that case. These costs will vary according to the nature, quantity and location of the dredged material.

It is important to note that the costs associated with use of the material will almost invariably be greater than those associated with sea dumping, although the additional costs associated with use need not be incurred by the authority responsible for dredging (as shown by the case study of the Harwich Haven Approach Channel in Box 8.7.1).

The fees associated with obtaining a MAFF sea disposal licence are reviewed annually to bring them in line with the estimated cost of running the licensing system during the fees year. The fee consists of two elements: an administrative fee and an assessment/monitoring fee. For the disposal of dredged material there are currently four charge bands which relate to the quantities of material to be disposed of. At present, the charges are in the range of £1000 to £7000. Information on licence fees relates to the year 1994/95 and was taken from *Fees for Licences to Deposit Materials at Sea and for Approval of Oil Dispersants: A Consultation Paper*, MAFF (1994).

Box 8.7.1 The use of dredged material

Harwich Haven Authority: In 1993, material dredged from the Harwich Haven Approach Channel was used by the NRA to build gravel embankments and for foreshore replenishment. Although the NRA did not pay Harwich Haven Authority (HHA) for these materials, it did fund the additional costs associated with using these disposal options in preference to the cheaper deep sea disposal option (costed at around £1.70/m³ - as dredged - for both dredging and disposal). In so doing, the NRA was able to make considerable savings by avoiding the need to pay for materials dredged from open waters (where costs are in the region of £6 to £10/m³ depending on royalty payments).

The additional costs associated with undertaking the re-use options are given below. It should be noted that these costs are relatively low as a result of extensive co-operation and liaison between HHA, the NRA and MAFF, over a period of three years, prior to dredging.

The gravel embankments were formed by pumping sand and gravel from the dredging ship to the shore. The costs associated with this option arose from the setting up of the pipeline and associated equipment (£75,000 for 250,000 m³ of sand) and the cost of pumping materials from ships of 8000 m³ capacity to the shore (£1.40/m³ - as measured at the shore). It should be noted that these pumping costs would be reduced for sand alone and increased (considerably) for gravel alone. Foreshore replenishment involved spraying (or "rainbowing") sand onto the area around the low water mark from ships of 800 m³ capacity. The costs associated with this option were £2.50/m³ - as measured in the ship. For both options, the dredging ships which were used were different (i.e. smaller, shallower draft) from those which would have been used for deep sea disposal and this is reflected in the costs given.

In addition to these costs incurred by the NRA (including survey costs), HHA incurred costs related to the Environmental Assessment, site investigations and (pre- and post-) monitoring.

Port of London Authority: Material dredged from the River Thames on behalf of the Port of London Authority is almost always used for fill contracts. The material is fine, silty sand and is typically used in the development of marshland. Past uses include use on land for the Thamesmead prison site at Woolwich and for the Dartford Freight Terminal. The possibility of some slight contamination means that some end uses are restricted.

The system of use is well established, with dredgers matching dredge and fill requirements and thus effectively acting as brokers between the Port of London Authority and the end users. As a result of this system, the Authority does not pay for dredging and, in fact, charges a royalty on the material (around £1/m³ - as measured in the hopper) and a small administration fee.

8.8 RE-USE OF MATERIAL

Dredged material may be used directly following dredging (e.g. by the NRA for bank strengthening) or may be re-used following some form of treatment or storage (see Chapter 6). Dredged material may have some commercial value for a number of uses, including road construction, landscaping or reclamation (of land not covered by a Waste Management Licensing exemption - see Section 8.4) and as construction materials. It is important to note that where material is used directly following dredging then it is not considered to be waste and does not enter the Waste Management Licensing system (see Section 3.2). However, material which is re-used (i.e. used following treatment or storage) does require a Waste Management Licence. Alternatively, it may be the case that selective dredging could be used to avoid the need for treatment, although this may not always be possible and may have higher associated dredging costs.

The costs associated with the re-use of dredged material could therefore include:

- any sampling and analysis costs to assess the feasibility of re-use
- any costs associated with identifying potential users
- any costs relating to a MAFF licence for beneficial use (see below)
- any transport, treatment and storage costs
- any revenue obtained.

The MAFF licence (introduced in 1993) for beneficial use (e.g. beach recharge and enhancement of mud-flats or salt marsh) is applicable to "operations involving the beneficial use of material, more than 50% of which has been derived from dredging activity and which would otherwise have been disposed of to sea". There are three charge bands which relate to the quantities of materials involved and, as for sea disposal licences, the fee consists of two elements: an administrative fee and an assessment/monitoring fee.

Dredged material which essentially consists of uncontaminated sand or gravel is most likely to be of commercial value. Box 8.7.1 presents two case study examples of the use of such material. Silts and muds may also be used, for example for saltmarsh or mudflat replenishment, although the utilisation of this material is a relatively new practice (English Nature, 1993b).

8.9 IDENTIFICATION OF THE LEAST-COST DISPOSAL OPTION

It is apparent from the above that the costs associated with the disposal of dredged material can be broken down into a number of components. By way of an overview, these are compared for each disposal option in Table 8.9.1.

For a given dredging operation there could be a range of possible disposal options which will have a number of different cost components. In some cases, it will be easy to identify the least-cost option. For example, where bankside disposal is an option, then this is likely to be the least-cost option when compared with other options which have transport and other associated costs.

For other dredging operations, however, the least-cost disposal option may not be so easy to identify and a number of cost estimates may need to be prepared. This may be the case even where there is only one possible type of disposal option. For example, consider a dredging operation in an area where the only available disposal sites are lagoons or landfills. In determining the least-cost method of disposal, the following "options" could be considered:

- disposal of wet material to the nearest site by transport in conventional lorries
- the use of specially prepared (but more expensive) lorries to increase the amount of material which can be transported per lorry
- dewatering material so that transport (and perhaps disposal) costs can be reduced accordingly.

Similarly, for materials which are contaminated, consideration might be given to whether the treatment of materials to reduce contaminant levels could result in reduced end-disposal costs and thus overall costs.

With respect to the design and operation of a landfill site, licence fees can be minimised by limiting the type and annual input of wastes to the site. However, this must be balanced against any losses in potential revenue from the use of the site by others.

For most dredging operations, cost will be the main factor of concern. However, in certain instances, the (positive or negative) environmental effects of disposal may be of equal (or greater) concern. Such effects can be taken into account in the decision-making process in a number of ways. At a minimum, they can simply be identified and described in qualitative terms and included in an impact matrix (similar in structure to Table 8.9.1) which compares disposal option with impact type. Alternatively, such effects can be included as parameters in a multi-attribute scoring and weighting system which allows options to be compared in terms of an overall measure of impact. At the most detailed and sophisticated level, such impacts can be quantified and valued (where possible) for inclusion in a cost-benefit analysis (CBA). Further general information on the inclusion of environmental effects in the decision-making process can be found in *Policy Appraisal and the Environment* (DoE, 1991) and a discussion of the application of CBA to decisions concerning the beneficial use of dredged material can be found in *Beneficial Uses of Dredged Material*, the Permanent International Association of Navigation Congresses (PIANC, 1992).

Table 8.9.1 Cost items associated with different disposal options

Cost	Agricultural land[1]	Reclamation/ improvement[1]	Bankside disposal[1]	Lagoon[2]	Landfill[2]	Sea disposal[2]	Use/ re-use[3]
Site Identification/ Feasibility	✓	✓	✓	✓	✓	✓	✓
Sampling/ Testing	✓	✓	✓	✓	✓	✓	✓
Agricultural/ Other Consultancy	✓	?	?	✓	✓	✓	
Site Design				✓	✓		
Exemption Registration	✓	✓	✓				
Post-Exemption Deposit Aftercare	✓	✓	✓				✓
Waste Licence				✓	✓		
NRA Discharge Consent		?		✓	✓		
MAFF Licence						✓	?
Planning Permission		?		✓			
Transport	?	?	?	✓	✓	✓	✓
Transfer	?	?	?	?	?	?	?
Storage Costs	?	?					✓
Treatment	?	?		?	?		?
Spreading Costs	✓	✓	✓				?
Compensation Payments	✓		?				
Land Purchase or Lease	?	?	?	✓	✓		
Site Construction				✓	✓		
Site Operation				✓	✓		
Revenues from Sales		✓					✓
WAMITAB Training				✓	✓		
Site Remediation/ Closure	?	?	?	✓	✓		
Post-closure/ Aftercare				✓	✓		

Key:

✓ indicates that the cost item is likely to be incurred
? indicates that the cost item may be incurred in some circumstances
(1) exempt
(2) licensed
(3) placement itself outside Waste Management Licensing Regulations although treatment or storage may require a licence.

References

ARIMSHAW, R., BARDOS, R.P., DUNN, R.M., HILL, J.M., PEARL, M., RAMPLING, T. and WOOD, P.A. (1992)
Review of innovative contaminated soil clean-up processes
Laboratory Report LR819 (MR)
Warren Spring Laboratory (Stevenage)

BERKSHIRE COUNTY COUNCIL (1994)
Deposit draft - waste local plan for Berkshire
Babtie Public Service Division (Royal County of Berkshire)

BRITISH STANDARDS INSTITUTION (1981)
Site investigations
BS 5930

BRITISH WATERWAYS (1992)
National Sediment Sampling Scheme. Report on the sediment quality in British Waterways canals and navigations
British Waterways (Gloucester)

BROOKE, J.S. and WHITTLE, I.R. (1990)
The role of environmental assessment in the design and construction of flood defence works
IWEM 90 Conference (Glasgow)

BUTTERWORTH, J.S. (1991)
Methane, carbon dioxide and the development of contaminated land sites
In: *Methane. Facing the Problems: Second symposium and exhibition*
Charles Haswell and Partners Ltd., Wimpey Environmental Ltd., and Department of Mining Engineering (Nottingham University)

CIRIA (1993a)
A guide to safe working practices for contaminated sites
by J.E. Steeds, E. Shepherd, and D.L. Barry
CIRIA (London)

CIRIA (1993b)
A guide to the control of substances hazardous to health in design and construction
CIRIA Report 125
Thomas Telford (London)

CIRIA (1994)
Environmental Assessment: a guide to the identification, evaluation and mitigation of environmental issues in construction schemes
SP96
CIRIA (London)

CIRIA (1995)
Remedial treatment for contaminated land
SP101 to SP112
CIRIA (London)

CSITI, A. (1993)
Land disposal of contaminated dredged material and related issues: the state of the art review
Terra et Aqua (No. 53)
4 to 15

DoE (1986)
Landfilling wastes: a technical memorandum on the legislation, assessment and design, development, operation, restoration and disposal of difficult wastes to landfill including the control of landfill gas, economics, a bibliography and glossary of terms
Waste Management Paper No. 26
HMSO

DoE (1989a)
Environmental Assessment. A guide to the procedures
HMSO

DoE (1989b)
Code of practice for agricultural use of sewage sludge
HMSO

DoE (1991)
Policy appraisal and the environment
HMSO

DoE (1994a)
Environmental protection. The Waste Management Licensing Regulations 1994
HMSO

DoE (1994b)
Licensing of waste management facilities. Guidance on the drafting of Waste Management Licences
Waste Management Paper No. 4
HMSO

DoE (1994c)
Landfill completion: a technical memorandum providing guidance on assessing the completion of licensed landfill sites
Waste Management Paper No. 26A
HMSO

DoE (1994d)
Landfill gas: a technical memorandum providing guidance on the monitoring and control of landfill gas
Waste Management Paper No. 27
HMSO

DoE (1994e)
Environmental Protection Act 1990: Part II Waste Management Licensing. The Framework Directive on Waste
HMSO

DoE (1994; 1995)
Circular 11/94; Circular 6/95. Joint Circular from the Department of the Environment, Welsh Office and Scottish Office Environment Department. Environmental Protection Act 1990: Part II Waste Management Licensing. The Framework Directive on Waste
HMSO

DoE (1995a)
Landfill design, construction and operational practice
Waste Management Paper No. 26B
HMSO

DoE (1995b)
Development of a national waste classification scheme. Stage II: a system for classifying wastes
Consultation Draft
HMSO

DoE (1995c)
Waste Management Licensing (Fees and Charges) Scheme 1995
HMSO

DIN (1984)
Schlamm und Sedimente S4 - Elution von Schlämmen
Deutsche Einheitsverfahren zur Wasser - und Schlammuntersuchung

ENGLISH NATURE (1993a)
A natural way of defining our land areas
English Nature Magazine (No. 7, May 1993)
English Nature, Peterborough

ENGLISH NATURE (1993b)
Strategy for the sustainable use of England's estuaries
English Nature, Peterborough

GAMBRELL, R.P. KHALID, R.A. and PATRICK, W.H. (1987)
Disposal alternatives for contaminated dredged material as a management tool to minimise adverse environmental effects
Technical Report DS-78-8
US Army Engineer Waterways Experiment Station, Vicksburg, USA

GREATER MANCHESTER WRA (1994)
Conditions to be attached to Licence Number
PROFILE LICENCE PFILF3/REV: INERT LANDFILL - MINOR
Greater Manachester WRA, Revised 26 May, 1994

INTERDEPARTMENTAL COMMITTEE ON THE REDEVELOPMENT OF CONTAMINATED LAND (1987)
Guidance on the assessment and redevelopment of contaminated land
ICRCL Guidance Note, Second edition.
HMSO

KELLY, R.T. (1979)
Site investigation and materials problems
In: *Proceedings of the Chemical Industry Society of Conference on the Reclamation of Contaminated Land*
C.I.S. (Eastbourne)

LORD, D.W. (1987)
In: *Reclaiming contaminated land*
Edited by T. Cairney
Blackie (London)
62 to 113

MAFF (1991)
Code of good agricultural practice for the protection of water
MAFF

MAFF (1993)
Code of good agricultural practice for the protection of soil
HMSO

MAFF (1994)
Fees for licences to deposit materials at sea and for approval of oil dispersants: A consultation paper
MAFF

NRA (1991)
Proposed scheme of charges in respect of applications and consents for discharges to controlled waters
NRA (Bristol)

NRA (1992)
Policy and practice for the protection of groundwater
NRA (Bristol)

NRA (1994)
Leaching tests for the assessment of contaminated land
NRA (Bristol)

OSLO COMMISSION (1993)
Guidelines for the management of dredged material
Convention for the Prevention of Marine Pollution by Dumping From Ships and Aircraft
Oslo Commission Report (London)

PIANC (1992)
Beneficial uses of dredged material. A practical guide
Report of PIANC Working Group 19
PIANC, Brussels

RIX, I. (1994)
Testing proficiency in soils analysis
Land Contamination and Reclamation Vol. 2 (No. 1)
4 to 6

SPAINE, P.A., LLOPIS, J.L. and PERRIER, E.R. (1978)
Guidance for land improvement using dredged material, Synthesis report
Technical Report DS-78-21
US Army Engineer Waterways Experiment Station, Vicksburg, USA

STANDING COMMITTEE OF ANALYSTS (1981)
Methods for the examination of waters and associated materials
HMSO

US ENVIRONMENTAL PROTECTION AGENCY AND US ARMY CORPS OF ENGINEERS (1991)
Evaluation of dredged material proposed for ocean disposal
Testing manual
USEPA and USACE (Washington D.C.)

US ENVIRONMENTAL PROTECTION AGENCY (1994)
ARCS Remediation guidance document
EPA 905-B94-003
Great Lakes National Program Office (Chicago)

US ENVIRONMENTAL PROTECTION AGENCY AND US ARMY CORPS OF ENGINEERS (1994)
Evaluation of dredged material proposed for discharge in waters of the U.S. (Draft)
USEPA and USACE (Washington D.C.)

WARREN, R.S. (1990)
Chemical site investigations
In: *Contaminated land and waste disposal: Proceedings*
Ove Arup Partnership (London)

Further reading

BOARD, P. (1993)
Quality assurance principles in contaminated land analysis
Land Contamination and Reclamation Vol. 1 (No. 6)
201 to 206

BOWEN (1979)
Environmental chemistry of the elements
Academic Press (London)

COOPER, D.H. (1987)
Field measurement of analysis in maintenance dredging
In: *Proceedings of the Institution of Civil Engineers' Conference on Maintenance Dredging*
Thomas Telford (Bristol)

DELWEL (1990)
Shaping the environment: aquatic pollution and dredging in the European Community
Edited by M. Donze (published at the occasion of the eleventh lustrum of the Association of Dutch Dredging Contractors)
Delwel (The Hague)

FAILEY, R.A. and SCRIVENS, A.J.
Contaminated land: assessment and redevelopment business and the environment practitioner series
Edited by R. Hillary
Technical Communications (Publishing) Ltd.

US ARMY CORPS OF ENGINEERS WATERWAYS EXPERIMENT STATION (1991)
Management plan for the disposal of contaminated material in the Craney Island Dredged Material Management Area
by T.E. Myers, A.C., Gibson, EA.A. Daqvdeau, Jr. P.R. Schroeder and T.D. Stark
Technical Report DS-78-21 (Washington D.C.)